Robot Futures

Robot Futures

Illah Reza Nourbakhsh

The MIT Press
Cambridge, Massachusetts
London, England

MIT Press books may be purchased at special quantity discounts for business or sales promotional use. For information, please email special_sales@mitpress.mit.edu or write to Special Sales Department, The MIT Press, 55 Hayward Street, Cambridge, MA 02142.

This book was set in Stone Sans and Stone Serif by the MIT Press. Printed and bound in the United States of America.

Library of Congress Cataloging-in-Publication Data

Nourbakhsh, Illah Reza, 1970–
Robot futures / Illah Reza Nourbakhsh.
    p.   cm.
Includes bibliographical references and index.
ISBN 978-0-262-01862-3 (hardcover : alk. paper)
1. Robotics—Popular works. 2. Technological forecasting—Popular works. 1. Title

TJ211.15.N68   2013
629.8′92—dc23
2012024598

10 9 8 7 6 5 4 3 2 1

To Marti, Mitra and Nikou: you illuminate my life.

Contents

Acknowledgments

Many people in my life have played critical roles in encouraging me to be inquisitive and providing me with the knowledge and habits of mind that shaped my intellectual identity. I am deeply grateful to all of them, although were it not for one particular mentor, I would never have become a roboticist. Professor Michael Genesereth at Stanford University introduced me to artificial intelligence and robotics, then encouraged me to join his research group. He convinced me to change my plans following college, continue university studies, and enter the PhD program at Stanford University, and he taught me a form of academic precision and sharpness in thought, and an appreciation of societal impact, that transformed me.

My peers, friends, and family have read early versions of this book and provided invaluable guidance. They have shaped both the style and the essential content of this work: Mark Bauman, Nonie Heystek, Steve Ketchpel, Ben Louw, Tom Lauwers, Marti Louw, Matt Mason, Ofer Matan, Farhad Noorbakhsh, Alex Norbash, P. W. Singer, and Holly Yanco. Not only did Mark Bauman and P. W. Singer improve this book through their early reviews; together with Fatemeh Zarghami, my mother, they are my role models. They set a gold standard for how one educates, communicates, and informs for positive social change.

Jim DeWolf at the MIT Press has enthusiastically supported this endeavor, navigating the publication process with patience and transparency. His efforts, and the willingness of the Press to publish a work that is critical of technology and its ramifications, speak volumes about their commitment to express strong ideas of all flavors to the public.

The life of a robotics professor is mostly consumed with teaching, research, and fundraising—so much so that I cannot imagine normally finding the time and space to properly nurture a book. Carnegie Mellon University and the Robotics Institute gave me that time and space by granting a sabbatical, allowing me to leave my university responsibilities behind and travel to a distant place. On the other side of that voyage, the University of the West of England's Science Communication Unit, led by Professor Alan Winfield, welcomed Marti Louw, my wife, and I with open arms into a dual sabbatical, providing an environment for research collaboration that was warm and stimulating during our stay in Bristol. Finally there is the one person who created the most important space for me to write daily, and who served as my most regular intellectual peer in evaluating and tuning the ideas in this book—Marti Louw. Her science communication expertise and her willingness to take on the societal impact of robotic technologies have made her the ideal collaborator.

Throughout this book I describe some of my past projects. The first-person narrative can imply that I invented and built these new systems singlehandedly, but nothing could be further from the truth. The CREATE lab's thirty members—researchers, educators, administrators, and students—are the real drivers behind every one of these projects. These individuals have dedication, social empathy, creative intelligence, and technical know-how, and together they make our audacious dreams of creative technology for social change come true.

Note Regarding Jacket Art

The jacket of this book features forty-one distinct versions of the color blue (see chapter 1). To create these separate colors, I visited the Couleurs Leroux factory in Joigny, France, where original pigments and oil are combined by hand to create art quality oil paints. The forty-one colors resulted from mixing six fundamental blue pigments from Leroux together in varying proportions with titanium white: bleu de Prusse, outremer, coeruleum, bleu cyanine, indigo, and bleu de cobalt. I am grateful to Leroux and to the artist, Francisca de Beurges Rosenthal, for expert guidance.

Preface

In 1977 I walked into the first run of *Star Wars* with my parents, not knowing what to expect. Just for context, we were there because *Herbie Goes to Monte Carlo* was sold out. Two hours later I was transformed, branded with images of C-3PO and R2-D2, robots among people. This is how my love affair with robots started, and it is also how an entire generation of robotics researchers, about my age, set their eyes on robots for life. I have participated in the past two decades of robotics research, where literally thousands of research groups across the planet have worked to close the gap between the promise of science fiction's robots and the reality of commercial robotics.

As is common in a field as multidisciplinary as robotics, my own career spans many different forms of robot innovation. I have worked to improve robots' basic capabilities—a new vision system for seeing the world three-dimensionally, a new strategy for navigating indoor spaces without getting lost (Nourbakhsh et al. 1997; Nourbakhsh, Powers, and Birchfield 1995). I have also participated in the creation of robots that we have deployed around the world: a seven-foot-tall tour guide that led visitors for four years through Dinosaur Hall at the Carnegie Museum of Natural History (Nourbakhsh et al. 1999); miniature programmable

Mars rovers installed in the National Air and Space Museum; the Exploratorium and the Japan World Expo (Nourbakhsh et al. 2006). But most of all I have applied new robotic technologies to interactive devices, imbuing new products with robotic powers: a pogo stick that launches the rider meters into the air (Brown et al. 2003); a vision system used by artists to make their art respond to the viewer (Rowe, Rosenberg, and Nourbakhsh 2002); a panorama robot that turns a regular camera into a billion-pixel documentary tool (Nourbakhsh et al. 2010); a messaging system that helps kindergarteners stay in touch with their parents; a smart electric car that local mechanics can make using used car parts (Brown et al. 2012); a robot-building kit that helps middle-school students build and program any robot out of craft materials (Hamner et al. 2008).

Robotic technologies seem magical because they are transformative. A product we simply use becomes something that sees us, hears us, and responds to our needs. Robotics makes the products around us more aware and more alive, a trend that will accelerate dramatically in the next decade. This is because the ambition of robotics is no longer limited to merely copying us—making walking, talking androids that are indistinguishable from humans. Robotics has grown up and grown out of that mold.

Modern robotics is about how anything can perceive the world, make sense of its surroundings, then act to push back on the world and make change. But never ask a roboticist what a robot is. The answer changes too quickly. By the time researchers finish their most recent debate on what is and what isn't a robot, the frontier moves on as whole new interaction technologies are born.

But there is one special quality of modern robotics that is very relevant to how our world is changing: robots are a new form of living glue between our physical world and the digital universe we have created. Robots have physical sensors and motors—they can operate in the real world just as well as any software program operates on the Internet. They will be embedded in our physical spaces—our sidewalks, bedrooms, and parks—and they will have minds of their own thanks to artificial intelligence (AI). Yet robots are also fully connected to the digital world—they are far better at navigating, sharing information, and participating in the online world than humans can ever be. We have invented a new species, part material and part digital, that will eventually have superhuman qualities in both worlds at once, and the question that remains is, how will we share our world with these new creatures, and how will this new ecology change who we are and how we act?

Already, robotic technologies are the living glue connecting the physical and digital all around us. They are running through forests as giant, military robo-dogs with visual optics that use the Internet to classify everything they see. Smartphones can guess what you are doing using built-in gyros and accelerometers, and can plot your path outdoors using the GPS satellite network and indoors using wireless antennas whose positions are shared in massive online libraries. When you ask Siri a question on the iPhone, your iPhone packages your voice and sends it online, then powerful shared servers in the digital world formulate an answer. Your digitized question travels thousands of miles, and your iPhone's brain includes not only the device you hold in your hand but the entire digital realm. Tiny, flying robots can buzz around a building, find an open window, and zip in to perch on a ledge, mapping the exterior and interior in real time. Those maps can be instantly published online; experiences are not local or transient in the robot-connected world; they are packaged, published, and digested before we humans have even blinked.

To begin to understand how robotics will change us, we need to understand key areas of robotics research and innovation. We take inspiration from humans, and so the first question roboticists ask is, what makes humans intelligent? We think of human intelligence as a quality that is living and interactive, embedded in the context of the world in which we function. Therefore intelligence depends on two things: being meaningfully connected to our environment, and having internal decision-making skills to consider our circumstances and then take action. The environmental connection is two-way, and we term the inputs as *perception* and the outputs back to the world as *action*. The internal decision making that transforms our senses about the world into deliberate action is *cognition*.

Perception is the ability to collect and interpret information about the world using sensors—digital cameras, sonar rangefinders, radar, light sensors, artificial skin, and many others. Perception is easy on the Internet because everything has a digital form—online sensors are easy to build, and the signals are easy to interpret. An online artificial intelligence can play video games on par with human visitors because it can see as well as humans can online. But robot perception in the real world means recreating the sublime physical and visual processing systems we have—feeling a firm handshake, recognizing faces, animals, textures, and fleeting smiles.

Action is the power to effect change in the world. For decades, robots have acted effectively in constrained situations such as automotive assembly lines. Robots are historically hard, heavy machines with powerful motors and little flexibility. An automotive assembly plant welding robot moves with great speed and precision, repeating the same complex motion thousands of times a day, all in a steel cage that is off-limits to humans, because the robot could thoughtlessly kill with a single blow. But acting in our social, human world means moving from the constraints of the factory floor to the dynamic, unpredictable world in which we raise our families. Instead of speed and power, social robots need elasticity, pliability, and gentleness of touch. This has motivated researchers to invent new types of motors with built-in springs, and new control systems to push a shopping cart or unscrew a jar of honey.

Cognition is the ability to reason, to make decisions about what to do next. Cognition is close to the traditional AI dream of thinking like a human: if a robot can sense the world through perception and change the world through action, then cognition is about making decisions about what to do next. It is the glue that connects perception to action, just as our brains absorb information using the five senses, then make decisions about how to behave next, connecting our senses to our muscles using reflexes and thought. Cognition is also the area in which robots veer away from how all natural animals operate, with local brains that must make decisions independently. Robots have effortless access to the digital world, and in that disembodied sphere there is both massive data and superhuman processing power. Every robotic decision can be informed by everything its shared network of robot brethren have encountered, and even the decision-making process itself is subject to outsourcing—a robot can use powerful online computing services so that its own circuitry can stay lightweight and power-efficient.

From a cognitive point of view, the robot we might encounter on a sidewalk is more unknowable than any animal. We will not be able to distinguish potential Borg from homebrew 'bot. Is it one physical pawn of a massive online intelligence reinforced by vast shared experiences and knowledge libraries, or is it a four-wheeled computer running a toy program written by the precocious preteen next door?

The three core strands of robotics research inquiry—perception, action, and cognition—do not proceed perfectly apace, nor are researchers succeeding in mimicking the diversity of human abilities all at once. Rather, our research is a ragged frontier that, in some cases, already exceeds human capabilities in peculiar ways and, in other cases, seems to be refusing every effort at advancement. We are not really on a straight path to the artificial human, but rather on the road to a strange stable of mechanical creatures that have both subhuman and superhuman qualities all jumbled together, and this near future is for us, not just for our descendants.

New research is yielding major innovations ever faster, and, what's more, human-level capability has turned out not to be a special stopping point from an engineering perspective. Consider action. Researchers have already built walking robots that can walk downhill using zero energy. Soon these robots will walk more efficiently than humans. Robots will take routes on Yosemite Valley's El Capitan that no human rock climber can ever match. One Carnegie Mellon project has created new materials that adhere to walls like Gecko feet. Their prototype robots effortlessly climb walls and, shortly, ceilings (Murphy, Kim, and Sitti 2009). In the case of perception, robots will see with greater detail, and they will see not only the light that our eyes register, but also the light signals detected by insects and birds. They will detect smaller movement further away, and one day they will see even better in the dark than a spotted owl and will navigate better than a bat.

A robot moving down the street will see in all directions, not simply in front of it like humans. If that robot is connected to a network of video cameras along the street, it will see everywhere on the street, from all angles, the entire time it walks. Imagine this scenario. A not-very-clever robot walking down the street will have access to entire synthesized views of the street— up and down, behind you, down the alley, around the corner— and be able to scroll back through time with perfect fidelity. As you approach this robot, it might be cognitively much dumber than you, but it knows far more about its surroundings than you do. It stops suddenly. What do you do? There is no *common ground* established between you and this robot, just the fact that you occupy the same sidewalk. The well-referenced concept of *grounding* in communication was presented by Herbert Clark to explain how even strangers can have a productive, social interaction by relying on a shared background of beliefs, assumptions, and group experiences that bring meaning to the few words they exchange (Clark 1996; Clark and Brennan 1991). A whole new robot species will have little in common with us in terms of beliefs or experiences, and so the basis of effective communication will be simply missing. What's more, along certain dimensions these robots will be far better informed.

One certainty is that humans will be inferior to robots in some ways. Is this really important, since we are already less competent than our computers at particular tasks such as calculating, spelling, and timekeeping? After all, none of this embodied computational prowess suggests that robots will write books that we want to read, or that the conversation we have with a robot will be affective and emotionally fulfilling. But robots will share our physical, social, and cultural spaces. Eventually, we will need to read what they write, we will have to interact with them to conduct our business transactions, and we will often mediate our friendships through them. We will even compete with them, in sports, at jobs, and in business. How will this change us?

I am not a social scientist, but as a roboticist I can predict possible futures for perception, action, and cognition break-throughs. With these predictions in hand, I can paint portraits of how such advances introduce us to new robot experiences *in the wild*, and how these experiences may change the ways we function in society. Each chapter in this book imagines an ever-further robot future in which underlying robot technologies have advanced and yielded new ways that we and robots share our common world. The chapters dwell, not on the technologies, but on the ways in which each possible future surfaces new human side effects, just like new pharmaceutical drugs. And, like drugs, the side effects that matter most reach well beyond the individual, influencing human activities and, ultimately, our culture.

The lay reader may also need an introduction to the upcoming technology innovations that will transform robots from laboratory experiments to consumer items. Embedded in chapter 2 ("Robot Smog") is a tutorial, "The Near Future Robot Primer," that I hope provides enough technical context and detail to begin imagining the pioneering robots of the 2030s.

Today most nonspecialists have little say in charting the role that robots will play in our lives. We are simply watching a new version of *Star Wars* scripted by research and business interests in real time, except that this script will become our actual world. Robotic technologies will infuse products all around us. Familiar devices will become more aware, more interactive and more proactive; and entirely new robot creatures will share our spaces, public and private, physical and digital.

The robot future will challenge our sense of privacy. It will redefine our assumptions about human autonomy and free will. As we face more intelligent robots, so we discover new forms of identity and machine intelligence. Our moral universe will be tested by robot cruelty and robot-human relations. Our sense of physical space and reach will expand thanks to robot proxies, just as our personal sense of self will be diluted into a broader and shallower digital-physical footprint. This book imagines successive milestones in robot evolution so that we can envision, discuss, and prepare for change, and so that we can influence how the robot future unfolds.

Note: I will track and blog major, relevant advances in robotics on this book's companion blog page: http://robotfutures.org.

1 New Mediocracy

Furniture Nation headquarters, chairman's office, Fayetteville, Arkansas, August 2030

"An overstock of ten million bogus plastic chairs with personal umbrellas. Are you kidding? They're not all blue like this, right?"

"They are. Part of the aquamarine pool accessory line. The hole in the chair has a molded flange that holds the umbrella. You can't sell the chair without the umbrella, or the other way around. Ten million."

"How much are we bleeding on stock?"

"Thirty thousand a month. The umbrellas are junk fabric, so we lose the principal if we don't sell them this year."

"Okay, so sell them. Why bring this to me, to the board level?"

"Because the adbot trolled the consumer ecosystems, and it says these five cities can absorb all ten million. We can get 10 percent conversion for all shoppers with network history, 3 percent for everyone else it targets with mobile interactives. This only takes four weeks virally, but look at what happens in this one neighborhood—and this is typical. So this family will buy one, and look, when they sit down on it they'll look around and see five other identical chairs in five neighbors' yards within a week. The walls are too low to hide the damn umbrellas. This happens in every target region. We just thought this would really weird people out, maybe cause a backlash."

"So this is a map of umbrella density? Why so uneven?"

"Interesting story. We looked into that. You remember the sunblock lotion coup in Portland a year ago?"

"Yeah?"

"That was the adbot doing experiments with word-of-mouth marketing through social networks. We had a sunblock overstock and the 'bot nailed a really good path—it drilled down on fear of cancer in school networks. Customized marketing to individual schoolteachers about skin cancer from sun exposure—and word of mouth from there to parents of the kids in the classrooms. These neighborhoods had high levels of concern about the medical system—government health care issues, local schooling—a perfect recipe. We had huge conversion rates. In cloudy December Portland. It was fantastic. Anyway the adbot is projecting the same strategy—skin cancer—for the umbrellas with the same teacher demographics."

"You try adding more borderline school neighborhoods and cities, spreading the umbrellas out more?"

"Basically we can get down to one neighbor umbrella if we add six major regions, but then the margin drops from 65 to 45 percent. Twenty million bucks."

"What if the adbot only markets to people whose neighbors didn't buy?"

"That works. We can spread them out evenly, but then it takes ten weeks because the computer has to track individual user decisions before hitting more teachers interactively."

"Okay—have you looked at drive-by? I mean, if we eat a hundred thousand bucks and do this slower, we can sell one umbrella to every block. But then, if you drive through the neighborhoods, will you notice the same umbrella at one home on every block, or is it hidden behind the house? Will people see the regular pattern?"

"We tried that. They're 80 percent hidden from the front, so the pattern's invisible."

"Fine. Take the two-month hit, have the adbot sell one to every block in the first five cities. Did you fire the purchasing agent who stuck us with these white elephants?"

"He quit. He did it on purpose and then he quit a week later. After approval, before delivery. Won't happen again—we replaced him with a robot."

• • • •

In business, the spoils of success follow when an organization finds the best possible match between what it can promise and what the consumer desires. There have always been two ways to narrow the gap between promise and desire: change what you offer to better match the customers' desires, or find a way to change customers so they want exactly what you offer. Either way, companies need deep insight into people's desires, and so consumer data collection has always been an important part of business practice. Today automated Internet and phone-based marketing surveys provide aggregate data about public opinion, and this data informs marketing and sales decisions about what to promise and what to produce.

But aggregate data are always inexact because they only provide a small set of samples, at best separating populations by demographics and categories. To understand human behavior and human desires more acutely, companies drill down to observe, track, and measure individual, representative consumers with hands-on focus group sessions and consumer behavior analyses. This strategy provides finer detail, but the magnifying glass is expensive and limits data collection to a narrow field of view: you must hope that the conclusions drawn from a handful of people apply to the entire market.

Sometimes it is cheaper to change consumers rather than change the product. To this end, companies do not just study people but use sophisticated marketing techniques to change people's opinions. Overt place-based advertisements are rhetorical tools created to change consumer desire directly, and more subtle campaigns such as product placement in movies and video games can all influence what people want.

It is particularly in the digital realm that marketing and sales have reached new heights of effectiveness. Companies have discovered the perfect recipes for convincing consumers to participate in new online worlds, then bind them emotionally to the values, social networks, and virtual products within. People spend real cash to buy cows in Zynga's *Farmville*; thousands have even spent more than $1,000 each on virtual robotic weapons in BigPoint's *Dark Orbit*.

The problem is that the digital world is spoiling corporations. They have discovered perfect playgrounds, where they can manufacture new worlds, characters, products—without retooling a single factory. They now realize how incredibly profitable it can be when they track everything the customer does; after all, tracking, measuring and discerning customer loyalty and demographics are easy and cheap in virtual worlds. The digital world has managed to addict companies to total information and total control as if they were a new drug. And the natural growth strategy is to expand their newfound power to an even bigger market: everyone in the physical world.

This is where robotic technologies come in: they bridge the digital-physical divide, enabling massive information gathering and control to pervade the real world. The Furniture Nation vignette imagines a future in which a computer system generates and experiments with multiple marketing campaigns for specific consumer demographics to get the best possible conversion from advertising to sales. Human marketing executives already do this today; the difference is that the computer will run millions of custom experiments simultaneously, and will expertly uncover strategies that are so effective that it just might be able to create unimagined desires—even for an unseemly personal umbrella chair that no one actually needs.

Discerning and shaping consumer desire starts with observing consumer behavior, but tracking behavior at scale in the physical world has always been time-consuming and expensive. Just tracking what television channels a family selects motivated Nielsen to design and manufacture *set meters* that the company installed in selected homes, meaning that a small sample of television viewers would need to represent everyone's viewing habits. To understand shopper behavior, stores with large marketing budgets hire professional firms that conduct *traffic analytics*, observing and visualizing customer flow, heat maps, and dwell times—how long shoppers spend staring and fingering merchandise.

The Internet, by comparison, offers such a narrow, digital window that tracking human behavior is far cheaper: every website collects detailed information about where a visitor comes from, how he navigates through the website, and when and how he leaves. *Web analytics* is the study of capturing and analyzing human behavior on the Internet: which links on the website are most used; what search terms at what frequencies yield sales; what model of computers do the best customers own; and which web browsers do they prefer? The answers to these questions are such powerful drivers of future profit that companies spend more than $700 million each year on web analytics.

Because the volume of traffic on the Internet is huge, and because the incremental cost of capturing more information from web visitors is tiny, analytics faces the challenge of massive data understanding in a way no focus group coordinator ever did. Analytics can yield literally hundreds of millions of data points—far too many for human intuition to make sense of the data. So in conjunction with the ability to store very big data about online behavior, researchers have developed strong tools for *data mining*, statistically evaluating correlations between many different types and sources of data to expose hidden patterns and connections. The patterns, in turn, predict human behavior and even hidden motivations. Originally, pioneers in data mining were funded by the Department of Defense and the Centers for Disease Control. Changes in drugstore buying habits throughout a city might identify the next epidemic, and unusual purchasing patterns can help flag potential terrorists.

But data mining turned out to be so productive that it has unearthed far more than superbugs and bomb-makers. Correlations between website visitor demographics, web search histories, current events, and conversions from visits to profits yield information that, thanks to these statistical number crunchers, becomes actionable intelligence for business decisions. The marketing improvements companies make based on this data, such as *landing-page tuning*, are all about *conversion maximization*: how can the Internet experience of the online visitor turn into profit for the company as comprehensively as possible?

Simple changes to a website can have a magical effect on profits, adding literally millions of dollars to the sales register: Market electric lawnmowers aggressively to each person who reads an electric car review. Provide coupons for an all-canvas shoe to everyone downloading a vegetarian Kitchari recipe. Two days into a bleak rainstorm predicted to last a week, raise prices on umbrellas but promise free next-day delivery.

In addition to the aggregate consumer data that every company now uses to better understand its customers, companies in 2001 began to create new technology that could observe the behavior of individual web users with greater fidelity than ever before. One of the pioneers, Vividence, designed *instrumented browsers* that capture every mouse and keyboard input in concert with the positions of scroll wheels and browser snapshots. The business model is straightforward: the analytics company recruits tens of thousands of volunteers who agree to use the instrumented browser to mediate their Internet experience. This means every mouse click, scroll, and keyboard stroke is recorded together with information about each user's demographic background and just where on the Internet she was before, during, and after shopping. Next, the analytics company sells its services to a large corporation that wants to know how to lift its conversion rate.

The volunteers are sent to the corporate site to book a travel ticket, and every detail about their experience is recorded, from the scrolling of pages to the speed with which they fill out web forms, click in and out of the company's website, and conduct any other activity on the Internet. The web's digital nature has greatly decreased the cost of recording each volunteer's precise online shopping behavior, yielding a treasure trove of information that can help the company fashion the most profitable user experience possible. It only takes massive data collection followed by very powerful data mining. Today, companies such as Keynote provide these monitoring and experience testing facilities as turnkey systems for any web or mobile phone product.

This form of voluntary instrumentation is no longer confined to the Internet. You can see examples throughout real-world shopping experiences in the form of customer loyalty cards. By creating perks such as gasoline or sales discounts, companies convince ever-larger proportions of shoppers to use unique identity cards, ensuring a full record of who buys what over the course of months and years, regardless of whether you paid with cash or credit. Marketing discoveries resulting from data mining on these national shopping records can even trigger in-store renovations. One particularly colorful urban legend still pervades marketing and sales lectures: a statistical correlation of beer and diaper sales from 5:00 p.m. to 7:00 p.m. suggested that beer displays in the diaper aisle would tempt fathers with newborns into impulse-buying beer on the way out during the afternoon emergency diaper run (Power 2002).

But what about privacy? Because loyalty cards are opt-in by default, we are all implicitly volunteering to give up privacy in return for saving money. But not all loss of privacy is opt-in. Internet shopping sites can glean details about every consumer even without explicit permission: website clicking behavior, purchasing habits, search terms that led to the shopping page, geographic locale, and computer details are all freely available. To stop this form of data collection, a shopper would need to use special *anonymizer* software to hide his computer details, and this is done by a tiny fraction of Internet users—less than four hundredths of 1 percent. Preserving privacy in modern times demands both a high level of technological literacy and a willingness to lose economically, or to pay more than fellow consumers.

The price-privacy bargain is likely to change even more rapidly thanks to major advances in robotic sensing. Consider a company that wants to go all the way, discovering the highest conversion rate possible for all potential customers everywhere.

The first challenge is that all possible customers are not just on the company's Internet landing page. To really maximize profit, companies need to take all the analytics and data mining tools that work so well on a landing page and extend them to every possible interaction the company can imagine. Robotic sensing will drive this very innovation. Landing page tuning will bust out of the Internet and become *interaction tuning*. Companies will apply their analytics engines to all interaction opportunities with people everywhere: online, in the car, in a supermarket aisle, on the sidewalk, and of course in your home.

Optimizing all consumption-driving interactions everywhere sounds like a tall order. To find the perfect triggers for every situation, a company would have to try out millions of strategies, finding the best ways to attract and convert contact into sales for every conceivable situation. Although this scale of experimentation is impossible in the physical world today, it is already common practice in the digital world. In web-based *A/B split testing*, businesses create competing website designs, then choose different designs for different customers who come to shop online. Over time, the system accumulates hundreds of thousands of visits to design A and design B, and all the aggregate statistics that are collected enable a true analysis for profitability.

In *multivariate* testing, more sophisticated data mining enables companies to make dozens of changes at a time to websites—designing, generating, and running experiments that simultaneously verify which ads, what wording, what fonts, even what colors maximize profits. Even very successful websites do this today, sending 99 percent of their traffic to a tried-and-true design, but risking 1 percent of the traffic on new variations, constantly experimenting on shoppers to discover ever-better conversion rates from visits to dollars. When Google was choosing the right shade of blue for a navigation bar, the company famously performed A/B split testing across forty-one different shades of blue to ensure that the final design would be absolutely unbeatable (Holson 2009). When numbers are large and hundreds of millions of people are in play, the tiniest improvement translates into breathtaking levels of profit.

On the Internet, the concept of active experimentation on shoppers can generalize broadly. Give custom discounts to customers shopping for the same object based on search history, location, or purchase history. Present different price comparisons based on search terms, and find all the right price-demographics strategies that maximize profits across the board. All of this works because the Internet has massive volume, it is instantly easy to modify, and it is simple to instrument since everything about your computer is already digital.

As robotic perception matures, dynamic marketing strategies, already in use by the smartest companies online, will become standard practice in the physical world. Suppose you are taking your family out for burgers and fries at the local fast-food chain. What if the store was watching you, and as you drove up the store-bot realizes it's you—the fellow who comes once a week with his family and almost always orders the same five sandwiches and two large fries. The store generates an order to the short-order cook, and the food is cooking by the time you have parked your car. The store achieves faster throughput because you do not wait, and the store throws away far less expired food, because it did not have to make a steady stream of fries no one purchases to keep customer lines short. It made fries when it knew, almost for certain, that you would buy them. Everyone is happier in this model, and the store improves efficiency and profit margins.

Now the kicker: this is not science fiction; it was demonstrated five years ago. *Hyperactive Bob*, a computer vision system tied to cameras around a store perimeter, watched the incoming cars (Shropshire 2006). After months of data mining on makes and models of cars and which orders correlate to each type of vehicle, the system reliably estimated what the short-order cooks should deliver as customers drove up. *Bob* extends the Internet landing page strategy to the parking lot. Even privacy advocates have trouble finding fault with *Bob*. The computer system is only recognizing a car and making a guess about what the car's occupants will order. If the company does not sell that information and does not associate the purchaser's identity with the car's details, then the invasion of privacy can seem minor. The company can even expunge details about *when* the car visits, removing information that otherwise could have legal value in a court case, for instance, establishing the validity of an alibi.

Recognizing cars well enough—*perception*—and making the right decisions about what to cook—*cognition*—were unthinkable at this level ten years ago. Today this is nearly standard practice. Researchers train video cameras on parking lots, fine-tuning computer vision software that recognizes pedestrians, measures their walking direction, and even learns their frequent destinations. Other researchers attach imaging systems to the four corners of a bus to scan the streets for bicyclists that might be in danger when a bus turns right. The vision system detects the cyclist's speed and direction of travel, and compares this to the bus driver's upcoming intentions, sounding an alarm if appropriate. Scientists have even developed computer algorithms and cameras that can focus on hands so precisely as to interpret sign language in a laboratory setting.

Projecting two decades forward, I can confidently say that tracking and understanding general human behavior in structured environments such as supermarkets will be largely a solved problem. Consider your behavior on a street, walking. A computer vision system will easily detect your gaze direction, how you walk, where you linger, what products you touch and try on, the exact set of places your eyes settle on during your entire time in and around a store, how excited you are as you talk to your friend about items, what expression your face makes when you turn the price tag over, what you buy, the cadence of your steps as you walk away, how old you look, and what your companions look like.

In November 2011, *Euclid Elements* emerged from stealth mode as a startup company dedicated to extending analytics from the Internet to physical stores (Perez 2011). The company's press release describes real-world equivalents of landing page and click-through statistics for the brick-and-mortar store: foot traffic, retention rates, dwell times, window conversion rates, and customer loyalty. Computer vision is not yet far enough along for broad facial analysis and tracking; instead, Euclid leverages the smartphone network, detecting and tracking each customer's position using unique smartphone Wi-Fi signatures. Cell phones are so ubiquitous that even the weakly opt-in nature of loyalty cards is unnecessary in this business proposition. Although insisting that personal identity is not recorded, Euclid explains that *for customers who still feel uncomfortable, stores will display a sign indicating how to opt-out of the data collection process.* This is the very beginning of the marriage of online and physical marketing intelligence techniques, and as sensing technologies progress, the boundaries of privacy will be regularly challenged anew.

In research labs today, facial tracking can measure which way you face and what you are looking at. Face interpretation software finds your eyebrows and your lips, and measures their shape to record your facial expressions. Eye analysis finds your pupils, estimates the angle of your eyes, and finds the objects of your gaze in the room. Moving facial expressions can already be recorded, analyzed, and transformed into estimates of emotional state—all automatically.

Now imagine the Nielsen set meter of the future. It not only detects your selected television channels but also how many of the family members are paying close attention to the television screen. It measures facial expressions, inferring emotional state as programs play out. Do you laugh at the jokes that are supposed to be funny? Do the commercials pull you in so that, as you stand up to take a break, you are instead drawn in by an outstanding ad? Does a show bore you and cause you to change the channel? Is a show so good that when your phone rings, you tell the caller to call back later and hang up? Or even better, do you glance at your cell phone during the show, check the caller's identity, and choose to ignore the call? Now break out of the living room. The set meter is all about advertising—it measures your reaction to advertisements in the bus, in your car, everywhere you walk where you might glance at an ad, and now the ad glances back at you.

This ability to computationally observe one person precisely, in the physical world, is just the tip of the iceberg. Computer vision systems will be able to do this for *every person, all the time*. In this robot future, web analytics invades our real world. Data mining will be up to the task. Human behavior, when enriched from mere mouse clicks to the tiniest facial gesture, will be a much harder problem in terms of both perception and data understanding, but two decades is ample time for exponential increases in the volume of data that machines can recognize and pick apart.

But what about the creation and manipulation of customer desire—what are the real-world equivalents of A/B split testing and customized online pricing? Companies are already proto-typing digital walls that will replace fixed advertising posters throughout physical stores (Müller et al. 2009). These digital walls will contain embedded computer vision systems that track face and eye movement, giving them direct access to knowledge about who is looking at the wall. Computer vision will not only track fine-grained human behavior, but will also be able to esti-mate age, sex, even fashion sense. Spoken language accents will yield clues about each customer's socioeconomic class, ethnic-ity, and educational level. Since the digital wall can change con-tent in an instant, this means that every advertisement will start to observe and individually experiment on consumer behavior.

Stop and think about this for a moment. The ad can observe deeply and it can customize interactive media to suit any experi-ment the adbot wishes to run. Granted, a particular person may not be highly predictable. But when the ad sees everyone doing everything, and can try every possible strategy out, then over time it will find the ways in which every category of person can be manipulated. Reimagine the customized advertising scenes from the 2002 film *Minority Report* with this level of personal optimization in mind, and the ads feel less like hapless clutter and more like a deadly accurate manipulation of desire.

The interactive ad will know how to push everyone's buttons, and what those buttons' labels are for various classes of people. This is an odd new form of human remote control. The button labels will be different for every demographic, but nonetheless there it is, in the hands of business entities. They know the space of possible ways in which they can change desire and thereby change behavior. The people's universal remote. Taken to the large scale, this becomes a bottom-up way to influence the behavior of society.

There is a hyper-rational argument that suggests this is a win-win situation. If the world can customize its messages just to me, and convince me that I really want a lumbar cushion, then I buy that cushion and the world convinces me that I am truly happy with my lumbar cushion, isn't that great for everyone? After all, I'm genuinely happy and the company is profitable. One month ago, I heard these very words from my friend after I described the conclusions put forth by this chapter. If only consumption and contentment defined the totality of the human condition. This reminds me of a scene in *The Matrix* when Cypher, a human rebel, strikes a deal with Agent Smith. He agrees to double-cross his friends in return for permanent, manufactured bliss: "You know, I know this steak doesn't exist. I know that when I put it in my mouth, the Matrix is telling my brain that it is juicy and delicious. After nine years, you know what I realize? Ignorance is bliss."

Perfectly manufactured desire is, of course, not only about bliss—we need to remember that at its foundation is business logic. The calculus of success has always been about closing the gap between promise and desire. But when desire can be so directly manipulated, individually, then business decisions may change more radically yet. Do not worry about *promise* nearly as much. Instead consider what consumer demands are in the best interests of the company, within the range of feasible desires that the company's universal remote can manipulate, and then follow through. The desire of people—their entire purchasing trajectory—becomes deeply influenced by the idiosyncrasies of which raw materials are cheap, which overstocks the company has, and which currency exchange rates happen to be favorable this month.

In the context of today's presumptions about personal privacy, this level of consumer profiling, model building, and customized interaction may seem unacceptable. But the transition to this scenario does not happen overnight, and as we have seen privacy eroded for economic gain for years, so there is every reason we can expect this trend to continue. Much of the marketing optimization outlined in this chapter can be performed without associating actual identities with people—just by aggregating across types of people using estimates such as age, gender, and profession. Location presence and personal contact information can be treated as an opt-in just as it is today, and a great majority will happily continue to sign up in return for discounts and special treatment. Economics and technology are powerful forces—far more compelling in societal terms than the desire for personal privacy.

The concept of massive experimentation and data mining to yield perfectly manufactured desire also extends beyond business, notably to the political process. Politicians, like business leaders, use surveys and focus groups to try to understand their constituency, then formulate rhetorically compelling arguments that reflect the constituency's interests in what they offer. Already, political action groups use targeted, just-in-time marketing at the very edge of technology, with microtrends, robo-calling, and robo-SMS campaigns for specific neighborhoods and demographics. The story is nearly the same as selling a product or service. But when a political action committee is able to measure the emotional response people feel to campaign ads, and is able to experiment at the person-by-person level, then it will create a model, thanks to data mining, that predicts how the populace will respond to alternative, customized campaign messages. In this case the universal remote does not influence what we buy, but who we support.

This approach makes today's autocratic despot look downright coarse, with his television station and radio station that blare one-size-fits-all propaganda to an entire populace. In this robot future, he will unleash interactive, custom messaging to every citizen that has just the intended effect. Democracy depends upon informed choice, which intends to invest the populace with a role in its future representation. When desire becomes so perfectly manipulated, thanks to well-honed, automatic, interactive experimentation and data mining, then the populace is no longer making an authentic choice at all. Instead, interactive new media is literally replacing independent human judgment. The title of this chapter, "New Mediocracy," denotes a possible robot future in which democracy is effectively displaced by universal remote control through automatically customized new media.

2 Robot Smog

Senate Subcommittee on Waste Disposal & Public Safety,
Washington, D.C., April 2040

[partial transcript]

MR. HOBSON: But it was essentially your invention, was it not, Mr. Lamb? You are the founder as well as the present CEO.

MR LAMB: Yes, but there are many similar robot toys on the market, Senator. I designed botigami when my peers were also manufacturing similar flying toys.

MR. HOBSON: It seems to me your product was, and still is, different. Every other flying toy we found for sale uses batteries. Isn't yours the only one in this category that uses Solarflex material?

MR. LAMB: We have not done extensive research into how all the other products work.

MR. HOBSON: If you used batteries, they would run out of power and we wouldn't be having this congressional inquiry, would we? So, tell us why you chose to use Solarflex.

MR. LAMB: It is far more suitable for flying robots. The solar material is so thin and efficient that it's lighter than batteries for the amount of power our 'bots need, and we were able to make the foldable structure of the kit out of the solar panel material. When the child purchases the botigami, he can open the box and begin playing right away—there is no need to buy batteries or fuel cells, or get the right tool and open a battery compartment. It is a more elegant solution by every measure—for me as an engineer.

MR. HOBSON: But batteries naturally limit the lifespan of an electronic object, or robot for that matter. Did you think about the cradle-to-grave issue when you chose solar power?

MR. LAMB: Actually photovoltaics is better because the child's play is not interrupted by dying batteries. And when the sun sets, the botigami stops—so we did not need any on/off button this way. That reduces manufacturing cost.

MR. HOBSON: We will come back to the on/off button question. But let's finish comparing it to the other flying robots out there. What is your share of the total market?

MR. LAMB: We have roughly 85 percent of the net sales volume. In terms of units.

MR. HOBSON: And how many botigami kits have been sold in the past five years?

MR. LAMB: Just under one hundred million internationally. About half in the United States.

MR. HOBSON: So, just to reiterate, do any other competing products have solar power and no on/off switches?

MR. LAMB: I'm sorry I can't answer that. I have not done that kind of competitive analysis. But my marketing team can do a survey and get back to you.

MR. HOBSON: Now, a second feature I want to call out: how about the gaze tracking? Do you know of any competing products that have the eye contact behavior built in?

MR. LAMB: I don't know of any, no.

MR. HOBSON: Why did you do it?—the eye contact swarm behavior—why design that in?

MR. LAMB: Have you read Gabriel García Márquez? *One Hundred Years of Solitude*?

MR. HOBSON: No.

MR. LAMB: It is a beautiful piece of fiction steeped in magical realism, and there is a scene where butterflies cluster around the protagonist in a breathtaking way. The eye contact interaction is an elegant way to recreate that experience for children. If you fold and set up five botigamis in a row along the wall, then look at each one in turn, they take off and flutter around your head. Soon you have five robots fluttering around you. Besides, the all-in-one vision chips do eye detection and eye gaze tracking right out of the box, with very little additional weight and almost no power consumption. Adding this to the botigami makes the robot much more fun, with very little change in BOM cost.

MR. HOBSON: BOM?

MR. LAMB: Bill of materials. Total cost of all the parts in the robot.

MR. HOBSON: If this eye contact feature is so great, why don't your competitors have this built in?

MR. LAMB: I don't know how they chose the play patterns they chose.

MR. HOBSON: What do they do instead? Surely you know that.

MR. LAMB: The one or two I have looked at make use of remote control. Direct control by the child.

MR. HOBSON: So, just to be clear: you did eye contact because you wanted to recreate an experience from a book.

MR. LAMB: And we hit on a good design. I mean, it makes for a compelling interactive experience. Clearly the consumers love it.

MR. HOBSON: Interactive experience? Okay. Please look at this picture, Mr. Lamb. Do you notice how many people are wearing very dark sunglasses in Central Park here? These are from the surveillance cameras, taken last week. Here, please note exhibit 5 and put this on display, clerk.

MR. LAMB: I see the sunglasses, sir. Yes.

MR. HOBSON: Okay. How about the people without sunglasses on. What are they doing? Describe it for the transcript.

MR. LAMB: They seem to be looking down at the sidewalk.

MR. HOBSON: No. They are stooped over, looking at their feet as they walk, Mr. Lamb. Here. Here is a time sequence from six years ago. One year before you invented botigami. Same park. Same time of year. Cloudy day. Notice the difference? Everyone is walking upright. Talking to each other. Laughing. Relaxed. Do you, Mr. Lamb, notice the difference when you yourself are out, walking in the park?

MR. LAMB: Senator, I am running a very large corporation. Unfortunately, I do not have time for a stroll in the park.

MR. HOBSON: Well, let me tell you what the sidewalks are like, since you do not go there. Everyone is afraid of being spotted by one of your robots, and making eye contact with it. Because off they fly and start circling round and round. So everyone avoids eye contact with nearly everything, since there can be a botigami jammed up anywhere—wherever it fell when the sun set the day before. Nobody looks around anymore, Mr. Lamb. In five years you've singlehandedly changed how people stroll through the park, with a $30 child's toy. My time has expired and I will recognize Mr. Remus.

MR. REMUS: Thank you. Let's turn to the public safety concern of this inquiry. Mr. Lamb, you are aware of how individuals have been hacking botigamis to create new creatures?

MR. LAMB: I am aware of some efforts in that vein, Senator, yes.

MR. REMUS: My nephew was playing at St. Andrew's Park, in Bristol. And he looked at what was apparently an exercise robot dog for runners. But it was actually a botigami spliced to an exercise robot. This 100-pound machine came running toward him at ten miles an hour when he made eye contact. As tall as him, it circled around him for twenty minutes, trapping the poor child. Are you aware of this specific hack?

MR. LAMB: I had not heard of that particular one, no.

MR. REMUS: Well I took a look on the Internet. There is an online instructable for taking your botigami's vision processor and replacing exercise 'bot controllers with your botigami controller. Apparently the similar architecture of the two machines makes this a mere one-hour operation. Is this possible?

MR. LAMB: Most robot devices use very similar parts because we purchase the same basic sensing and computing components from the same manufacturers. It is natural that some parts are interchangeable.

MR. REMUS: Well, public safety can suffer from that. Can you make the next version of botigami impossible to hack that way?

MR. LAMB: Impossible to hack? I don't understand.

MR. REMUS: I'll make it plain—botigami is a major public nuisance, and now it looks as if it may even endanger the public. We want to insist that you make the next version unhackable.

MR. LAMB: But that's not possible. If you build something—any robot—it can be reverse-engineered by definition. Nothing is hackable or not hackable. Everything can be modified. They're just built products, after all, not biological creatures.

MR. REMUS: That's absurd. That's like telling me that every time you invent something new, it will make a mess of our city all over again because someone will figure out a terrible use for it. In the twentieth century, didn't we have decades of invention that did not result in unintended disasters? Have you even looked at the land-fill problem we have now? Do you know how much maintenance prices have shot up? Every time we touch earth, we dig up some botigami, it charges on sunlight, and off it goes looking for human eye contact. Have you seen pictures of what waste disposal sites look like now, Mr. Lamb? We do almost all operations at night, just to avoid waking the robots. What on earth do you suggest we do, Lamb, if every robot is always hackable?

• • • •

In 2004 Rufus Terrill, an ex-marine, purchased a bar in downtown Atlanta, renamed it *O'Terrill's*, and set out to run an Irish-themed pub in the United States. The pub was located near both luxury apartments and a homeless shelter, and homeless vagrants fre-quently sat on the sidewalk by the pub. Terrill was upset by the environment of vagrancy and drug activity around his establish-ment, and he built a robot in 2008 to help clean his sidewalk.

The robot's chassis was a three-wheel electric scooter; the torso a meat-smoking barbecue. Bright red lights on the robot were the taillights of a 1987 Chevrolet Camaro. A home-alarm loudspeaker system mated to a walkie-talkie provided the robot with Terrill's voice. The robot had a moveable turret, including a bright spotlight and a high-pressure water cannon.

O'Terrill's Pub is now signposted as "Home of the BumBot." Terrill sells BumBot-emblazoned T-shirts both online and at the pub. This may be the first local pub where a remote control robot keeps the area clean of unwanted people through intimi-dation and threats.

Two trends have made a robot such as BumBot—and the recep-tion it has received—possible. First, the cost and complexity of technology needed to make a custom, remote control robot has decreased significantly. Almost anyone can, in a weekend, make a primitive robot, and the construction kits now available make that first robot larger, heavier, and stronger. The second trend is one of attitude. We, as a culture, celebrate scrappy do-it-yourself (DIY) invention, through the world of Burning Man, Maker Faire, *Craft*, and other such outlets. By building a robot out of a variety of found parts, Terrill joined a celebrated club of enthusiasts who appreciate the idea of building a new device through *remix*—recycling and repurposing by using parts in ways they were never originally intended to be used. Even with a purpose as ethically troubling as the BumBot's, the aura of Terrill being a modern-day inventor provides an afterglow that softens the concerns many have regarding the actual details of what he does with the robot.

Over the next couple of decades, advances will transform DIY robotics, with the devices increasing in sophistication, diversity, and, most important, the degree to which they inhabit the same spaces we do. There will be millions of Terrills, and each one, with a unique agenda, moral compass, and vision, will make human-sized creations that embody his dreams in steel and bytes. This public democracy of built machines will be omnipresent and massively diverse. Could such a spectacular zoo imprison us? Smog is a portmanteau that combines the natural and the artificial; fog simply reduces visibility, but when smoke and haze mix together, then quality of life decreases: runners cough, tennis players' lungs burn, and asthma cases in children bloom. Robot smog is a technological portmanteau: the visions of people, usually communicated through pen and voice, will soon adhere to the sidewalks and atmosphere of our physical world in literally invasive forms. There is a danger that we might suffocate in this world of dreams gone real. Could commercial robots such as solar-powered toys become both so autonomous and so easy to modify that we find our parks infested with robot inventions that will not leave us alone?

The history of the Internet contains a similar, albeit highly constrained, example that parallels the concept of robot smog. By the late 1990s, everyone wanted to author their own website. But creating a new website meant writing in HTML, a type of computer specification for websites, and this in turn demanded knowledge of computer languages with a sophistication originally reserved for computer programmers. This changed abruptly with the advent of Macromedia Flash, hand in hand with interactive software, such as FlashToGo, designed so that nonprogrammers could prototype and publish entire websites with no prior graphic design or computer programming experience.

The new tools catalyzed a great increase in the number of websites published every week, and many of them were filled with animated widgets blazing away thanks to Flash: butterflies flapping, hearts beating, words blinking, clip art breezing across the screen. Of course, the results were aesthetically unappealing, with clichéd, poorly designed, and annoyingly loud websites germinated and rapidly reproduced in a few short years. But the downstream effects of this new, easy publication process were far more profound. The Internet became everyman's stage. Like a soapbox in the town square but with a megaphone that reaches the ears of millions of neighbors, every person with an hour of time was empowered to espouse her own world view, her own personal politics online. This, in turn, led to a Balkanization of the Internet as more and more extreme ideas attracted the readership of like-minded souls, and the ethics of moderation and compromise lost out to extremes, exclusion, and exaggeration.

The parallel to robotics is straightforward. Technologies will make it far easier for anyone to make a custom robot over the next decade. Will this result in a zoo of obnoxious, exotic new creatures? The early explosion of animated websites was confined to the Internet. You could stand up and walk away from a loud website by simply leaving your desk. Even today's creative explosion of smartphone applications is limited in scope—they are inside your phone, and you can choose just which ones you download and use. But future robotic creations will be physically omnipresent. When your neighbor down the street makes it and sets it free, you may have to wrestle it out of your vegetable garden the next day. In this robot future, personal opinions are not just communicated; they are acted out by chaotic ecologies of robot minions.

To imagine what public DIY robot invention may produce in twenty years, we first need a robotics primer that describes the system improvements and robot breakthroughs that are likely and easily accessible in the near future.

The Near Future Robot Primer

I organize future robot breakthroughs into six categories of innovation: structure, hardware, electronics, software, connectivity, and control. The next six sections address each category in turn, forecasting two decades forward.

Primer 1: Structure

The structure of a robot—its chassis, geometry, and joints—is an underappreciated but extremely important aspect of robot design. The physics of how a robot can move, how much it weighs, and how that weight is distributed throughout its frame is essential for a successful design. Robots, after all, operate on the boundaries. We are often pushing motors just to their breaking point. The weight of an arm that has to be supported by a motorized shoulder directly impacts which motor technology is viable. This, in turn, dominates the final cost of the robot.

Geometry and weight matter so much that some research projects are dedicated solely to these issues. One such topic is passive walking. Our grandparents had wooden toys that would walk down a gentle slope without the use of batteries and, often, without even springs for temporary energy storage. Today's researchers have developed human-scale robotic machines that can walk down gentle slopes without any net energy expenditure. These pioneers study the boundaries of weight and mechanics to create machines that are so energy efficient that, when they do need to walk uphill, they will use just a fraction of the energy of conventional walking robots (McGeer 1990; Omer et al. 2009).

Doing research in robot structure used to be especially challenging because realizing new designs meant outfitting a research lab with high-end machining and fabrication equipment, including lathes, milling machines, and welding equipment. But over the past five years there has been a small revolution in low-cost, rapid prototyping thanks to 3D printing and laser cutting technologies. 3D printers can deposit hot plastic, one layer at a time, to create any three-dimensional form. Laser cutters can cut plastic, wood, and even metal into intricate patterns that can be assembled into a new robot frame. Every robot lab now has the ability to invent a new shape and have it prototyped and ready for testing in a matter of hours or days, at prices so affordable that they can make tens and even hundreds of experimental robot bodies in a matter of weeks.

The old days of robot research were marked by similar-looking robots populating every lab, since researchers purchased their robots from the three or four manufacturers who built the same, round trash bin–sized robots in quantity. Today, robotics labs have large collections of eclectic, custom-fabricated forms. The creative explosion is like the mammalian diversification of the Cambrian period: each lab is populated by robots of a variety of sizes, weights, and shapes—and each lab's robot "corral" looks nothing like that of a neighboring lab.

This robo-Cambrian explosion of diversity will especially benefit two specific areas of robot structure over the next ten years: high degree-of-freedom (DOF) robotics and flying robots. High DOF robots have large numbers of joints and motors, which means they can interact with the physical world in much more sophisticated ways than a classical robot-cart that can simply drive and steer. Degrees of freedom enable legs and hips, which empowers robots to use staircases and thereby access most living quarters; shoulders and arms access all of the devices that we humans use from day to day, from light switches to refrigerators to laundry machines. Even higher DOF's suggest even more hands, tails, and snake-level flexibility. A decade ago, robots were able to traverse monkey bars like a six-year-old. Today, snake robots designed for military surveillance can shimmy up a four-inch gap, and android-like home service robots can open a refrigerator door, see the bottles within, move the contents around, reach in, and take out the beer. We can easily imagine applications that exceed what humans can do, from scaling trees and running through a forest canopy to snaking through a sewer system, climbing any rain downspout, and searching through a disaster site by worming through rubble.

Snakes and other high-DOF robots inhabit a space where weight and motor strength are paramount issues, but these problems are even more critical in the case of flying robots. Rapid structural simulation, prototyping, and testing will be game-changing for flying robots. A visit to the local electronics store demonstrates some of the advances. A decade ago, small, electric remote control helicopters cost hundreds of dollars and flew for two minutes at a time. Today, anyone can buy and fly a palm-sized helicopter for fourteen dollars.

Robotic control of flying machines has taken this new generation of ultracheap helicopters to a whole new plane. Vision systems can track the exact orientation of the helicopter in real time, and automatic control systems can fly them with so much precision that the helicopters can swing up, turn sideways momentarily, and fly through narrow vertical openings with just an inch of clearance on each side. They can perch on a ledge, land to save battery power, and take off later (Mellinger, Michael, and Kumar 2010). Research laboratories have unveiled videos of twenty robot helicopters, hovering and flying in tight formation and even changing formation dynamically to go through an open window frame in synchronized pairs (A Swarm of Nano Quadrotors 2012). Today, these demonstrations depend on ceiling cameras and external computer control, but these limitations will soon disappear. Aerodynamics and control—robot shape, weight distribution, mechanical configuration, and automatic control—are the keys to flying robots that will be cheap, highly maneuverable, and able to stay aloft with very little power usage. Advanced 3D manufacturing, especially as new plastic and elastic materials become available, will make possible flying robots that are only dreams today. I believe the robo-Cambrian explosion is only now beginning, and the trajectory of inventiveness in scale and form will be steep.

Primer 2: Hardware

Hardware is the most difficult to predict category of robot technology development. Major innovation is absolutely necessary in terms of hardware, yet we cannot predict the pace of future hardware innovations by simply extrapolating from recent history. In computer design Moore's Law has reliably predicted the rate at which processors become faster, doubling in speed every eighteen months. But there is no Moore's Law for hardware technologies such as batteries and electric motors. In fact battery and motor research have witnessed multiyear droughts when hard-won research has failed to materialize into commercial success.

Yet advances have occurred, and have been game-changing when they have finally come. There is a legendary story that circles within the robotics research community regarding the Honda ASIMO, which was arguably one of the most telegenic charismatic megafauna to stoke the public's imagination from the world of robotics research. One of the lead Japanese engineers on the ASIMO project was on tour, demonstrating the robot and vaguely answering questions at universities across the United States. In an unscripted moment of candor, he said the one killer problem they had to surmount was the motor. That's it. Once they figured out how to make a motor for each of the joints in the robot, developing the rest of the humanoid was simple by comparison.

Motors really are a limiting technology impeding robots' advances. Part of this has to do with how differently motors work from the joints and muscles of biological systems. Our joints are lightweight and highly elastic. The flexibility of each joint and the stiffness of our muscles are controllable, and so you can parry in fencing while keeping your muscles relaxed, enabling the foil to move easily to the side; or you can arm-wrestle and stiffen the system up, making it rock solid and unmovable. Our joints are also extremely fast, to say nothing of hummingbirds and butterflies.

In contrast electric motors have historically been heavy, power-hungry, and stiff. If you add gearing to the electric motor so that it is strong, then it is also slow and hard—you cannot push on the motorized arm and move the gearbox backward. If you make a small robot arm with a built-in electric motor that is lightweight, it will be entirely unable to lift a coffee cup, even if the arm weighs as much as a human arm. And these are all simple examples compared to a robot that can lift an elderly person out of bed in a home and help him to the bathroom— activities requiring suppleness, elasticity, and great strength, all at the same time.

For a time many believed that the future of robotic motors belonged to systems that work like human muscles, and the muscle wire technology Nitinol was heralded as the next revolutionary step. Clever mechanisms were created that allowed this wire, which stretches and shrinks with the application of electricity, to work in parallel with more of its kind to carry weight and to extend its range of motion from millimeters to centimeters and more. However, challenges with overall energy efficiency and fatigue have narrowed the useful applications of such muscle wire to areas such as medical robotics, where a carefully controlled environment makes it nearly ideal.

Recent work in new motor designs, however, is bearing fruit in terms of highly dynamic, controllable motors. Research joints are being developed that combine electric motors with adjustable springs, providing an overall system that is able to cope with external forces much more elegantly. Ever more sophisticated motors now have force sensing built into them, with such fast control that the motor electronics can simulate any stiffness and elasticity desired by sensing how much pressure is on the motor externally and then responding to that pressure in real time. This will someday enable a robot hand to provide a firm but safe handshake, fold origami, and even crack an egg without breaking the yolk. One reason motors improve is that, even though motor technologies do not abide by Moore's law, they are benefiting from ever-improving onboard control electronics and software, and this gives Moore's Law a pathway to influence the future of motor technology.

Batteries are the other category of hardware that has the potential to seriously impact robotics' advancement. Making a revolutionary battery has proven elusive because, unlike the incremental progress in computer chip design that extrapolates forward years and decades, battery advances depend on fundamental, disruptive discoveries in the areas of chemicals and materials synthesis. Just when these discoveries will be made, no one knows. In computation, Moore's Law suggests a doubling of computer processing speed every eighteen months. In contrast, look at the advancement of battery technology. Lead-acid batteries were first created in 1860 and had approximately 30 watt-hours per kilogram (W·h/kg) of energy density, meaning that each kilogram of battery could provide 30 watts of power for up to one hour. This energy density value did not double until nickel–metal hydride (NiMH) came along in 1988, at 60 W·h/kg. The same machine would need only half a kilogram of NiMH batteries as compared to one kilogram of the earlier lead-acid chemistry. By 1997, lithium-polymer set a new level at 180 W·h/

kg, yet even today they cost much more to manufacture than the NiMH batteries of three decades ago. Now, for a shock, compare these figures with gasoline: 13,000 W·h/kg. A few drops of gasoline, 2.5 grams total weight, provides the same total amount of energy as an entire kilogram of lead-acid battery.

Here, in a nutshell, is where the challenge lies for electric cars and for small robots alike. To make this concrete, consider a 50-kg person-sized walking robot that needs 350 watts of power on average. To enable this robot to function for six hours at a stretch would take 70 kg of lead-acid batteries; 35 kg of NiMH batteries; 12 kg of high-end lithium batteries; or just ten tablespoons of gasoline.

There is plenty of business pressure to both improve batteries and to lower the power demand of mobile products. Thanks to the tablet and smartphone industries, these business investments will make future robots compute with fewer watts and they will eventually deliver lighter, cheaper batteries. But remember that battery materials often dominate the equations of cost—therefore economies of scale will not necessarily push the price of advanced lithium batteries down low enough for tiny, affordable flying robots to benefit meaningfully.

Just like motors, battery research will demand real innovation from engineers to produce a whole new generation of options. Alternatively, fuel cells and photovoltaics will need to make similarly revolutionary breakthroughs. All these technical challenges stand mainly in the way of small robots that need lightweight but highly energetic power supplies. In contrast, even incremental improvements in battery technology over the next ten years will be more than sufficient for bigger robots: those weighing tens and hundreds of pounds. These robots can comfortably depend on next generation fuel cells, large batteries, or even petrol engines to run, jump, and walk around town.

Primer 3: Electronics

Electronics trends in robotics have followed a circuitous path that only now has the sort of stable progress that illuminates the future. One of the first research robots was Shakey the Robot, built by the Artificial Intelligence Center of Stanford Research Institute (now called SRI International) in Menlo Park, California (Wilber 1972; Nilsson 1984). By 1971, this robot was already far ahead of its time: it could navigate cubicles in a research lab, visually identify its position, and recognize obstacles. Imagine, this was visual navigation through the use of video cameras at a time when the computer interface was still a teletype machine, not a computer monitor with text! The robot, Shakey, was not really confined to its six-foot-tall form but included room-sized PDP-10 and PDP-15 computers that constantly communicated with the physical hardware. The robot, in other words, was never self-sufficient, but depended on computing resources outside its own physical body.

The 1980s and 1990s saw a trend of researchers trying to achieve a sort of holy grail: the self-contained mobile robot. The challenge, as researchers saw it, was to stop depending on external resources as a crutch, and therefore to put all the resources that were required on board the robot. Given power limitations and weight limitations, this required significant electronics engineering to do as much of the perceptual processing as possible, and thanks to these efforts robots began to be entirely self-contained systems at the close of the twentieth century.

Then came the age of software-as-a-service, and its natural application to robotics, with the architectural desire to have the robot use as many Internet-based resources as possible. For instance, ask external resources to do the face recognition so that the robot can have minimal electronics on board yet be able to recognize every person in the robot lab and say good morning to them. The Internet, with its ever-increasing rates of data transfer, was seen as a solution to the problem of miniaturizing robots and yet retaining high IQ in their behavior.

Just as this return-to-Shakey philosophy was taking hold, along came the tangible side of video games. The Wii remote enabled video game players on the Nintendo PlayStation to physically swing at the video screen, and this meant that the video game electronics developers had started producing hand-held electronics that roboticists could borrow straightaway. Low-cost, accurate accelerometers and gyroscopes changed the game, because now one could densely pack a small mobile robot with local sensors that cost only dollars, thanks to the econo-mies of scale triggered first by video games and now also by smartphones.

This trend hit high gear more recently with the Microsoft Kinect—an electronic camera system designed so that human gestures and body movements can connect, through the Xbox, to video games. A tidal wave of research articles in robotics already announces ways in which mobile robots can now use programmable Kinect sensors to detect walls and obstacles and to enable human–robot interaction (HRI) in ways that were too difficult before—for instance, human sign language control of a robot or a dancing robot that uses Kinect to waltz with people. So we are returning again to the age of sophisticated onboard electronics, but what is different now is that high-volume, low-margin interactive consumer electronics have arrived. And with this renaissance in interactivity, every new electronics invention will have direct applicability to robots, and will be hacked as necessary to make robots smarter and better.

Consider the future of cell phones and games: from detecting whether you are in a meeting to sensing your position, environmental values such as temperature and humidity, whether you are sleepy or excited—all of these physically sensed electronics solutions will tumble into the robotics cabinet of parts. Every new way in which our cell phone ingratiates itself into our daily activities also makes robots able to detect and respond to our environment more comprehensively.

Of course, the lines between video games, online entertainment, real-world entertainment, and mobile communication technology will blur. All the devices in our sphere are becoming more interactive, and that interaction will be driven by more natural communicative acts, including the spoken word, eye contact, and gestures. More of our personal accessories will have personal initiative, acting on our desires or high-level instructions like a good concierge: working to find and book the right restaurant reservation, rearranging our day, even screening our phone calls based on observations of what we are doing.

Perhaps the word "robot" will become confusing since cell phones will behave more robotically, and robots will often be sophisticated telepresence communication devices, enabling us to tuck in a child or visit a work colleague when we are away. Come what may, what we think of today as robotic technology will be far more sophisticated and clever, thanks to ongoing advances in multiple electronically intimate industries.

Primer 4: Software

Software development in robotics revolves around a storied history of attempts at creating a standardized framework for programming that would, in theory, gradually raise the capabilities of all robots in a shared and open way. These efforts, usually extrapolating from the successes of standardization in computing, have taken the best practices of software development and have tried to apply them to robotics.

But in robotics such attempts have only been partially successful, stemming from the fact that robots are far more diverse than computers. Computers have similar architectures and very similar processors and parts. Robots can look similar on the outside but have entirely different guts: in one case with a radio transceiver and low-cost processor, in another case with a full-blown computer inside and custom sensors. Standardization across such diversity is nearly an oxymoron. Yet this has not stopped major corporations from doing their very best.

In 2002 Intel Corporation and others made a first push at developing a community and process for standardizing robotics (http://www.retf.info). Their idea was to model this effort on the successful community-based Internet Engineering Task Force (IETF), in which Intel was an important early participant. IETF succeeded in helping standardize the Internet by being inclusive, open, and consensus-based. Taking these concepts to robotics by proposing a Robotics Engineering Task Force (RETF) seemed like a rational first step, but the effort suffered setbacks as soon as the consensus-based approach was applied to even the simplest problem of all: generating a set of standardized terms for robot parts. Roboticists could not easily agree on definitions of *sensor*, *motor*, and *processor*—not because of lack of effort but because one researcher's motor is, literally, another person's sensor. Even the word *robot* was so hard to pin down that standardization seemed a goal whose time had not yet come.

Avoiding the hardware consensus strategy, Microsoft Corporation made its own effort to standardize the programming of robotics with Microsoft Robotics Studio starting in 2005 (Jackson 2007). Microsoft's goal was to create a unified programming interface for robots that would be valuable for hobbyists, educators, and industry alike. This application diversity forced Microsoft to grapple with a simultaneous demand for a shallow learning curve targeted at the hobbyist and educator, and for high-end performance such as real-time control for serious researchers and for commercial applications such as industrial assembly-line robots. At the same time, Microsoft wished to espouse a service-based architecture so that robots and robot processes could publish and subscribe to services such as face tracking, navigation, and remote control. In the end, satisfying the needs of such a diversity of customers has led to a product that is in use by some but has not become a broad community standard. It seems the year 2005 was still too early for a standard to be born.

In 2008, a new company, Willow Garage, threw its hat into the ring with the Robot Operating System (ROS). This product takes some inspiration from the success that Intel Corporation had achieved in the prior decade with the Intel OpenCV library, an open-source collection of computer vision routines that greatly influenced hobby, educational, and research computer vision work (Bradski and Kaehler 2008). By selecting important computer vision capabilities, then optimizing software for performing those skills on Intel computer chips, the company delivered highly competent vision behaviors into the hands of those who wished to create end user applications (Quigley et al. 2009). From ping-pong-playing robots to sketching arms that would photograph your face and make a line drawing, the

OpenCV library led to a blossoming of engaging demonstration programs throughout computer labs and science museums. With ROS, Willow Garage hoped to package key skills required for mobile robots to succeed, offering them all in a bundle that individuals would be able to use on Willow Garage's own robots and on other robots running ROS. These skills, including basic navigation, manipulation, tracking of moving objects, face recognition, and gesture recognition, have catalyzed a subcommunity of ROS researchers who are using, tweaking, and adding to the ROS repository communally.

While ROS has been particularly effective for high-end robots that have significant processing on board, other developments in the smartphone segment are promising equivalent advances in the arena of small, low-power robotics. Google Corporation's Android operating system is unleashing an architecture that allows smart phones to communicate directly with robots, and allows the phones to act as decision-making processors, or brains, for small robots. In the near future you may be able to build robots that communicate with the phone in your pocket and, through that phone, with the Internet at large. Your phone may even be a centralized brain, a queen bee, for the colony of low-cost robotic creatures buzzing around you.

The past ten years' experiences suggest that there will be no grand unified theory of software architecture for all of robotics. Our field is simply too diverse for that to make sense. However, major advances in services that all robots can use are already occurring, and the rate of such advances will only increase. Indoor navigation used to be a major research problem in robotics, with entire conference tracks dedicated to the question of how to keep robots from becoming lost. Today, entire indoor navigation packages are regularly used, as services, by undergraduates who need not learn about the theory of robot navigation to build a robot that can map and navigate a dormitory hall. In the tradition of Intel OpenCV and ROS, many more services will begin to appear, particularly as robots are able to go further and interact more richly with our physical world. Walking on sidewalks while sharing them with humans; dealing with elevators and negotiating entry and exit; cooking a simple omelet—these are all hard for robots today, but as soon as the hardware is in place, software solutions will wend their way into shared repositories where all robots can benefit from new capabilities.

No such behavior is more important than manipulating the built world. The ability to deal with the human home—picking up and moving cans, mugs, newspapers, and laundry—is absolutely critical to the robot's ability to play a productive residential role. Today, manipulation research is where indoor navigation was in the 1990s. Research pioneers are beginning to show off robots that can use a lightweight arm together with high-end cognition to see a table, analyze its contents, then maneuver the robot's hand to the table to pick up a desired object without knocking over everything else on the table (Srinivasa et al. 2010). These problems are surprisingly hard, involving reasoning about the shape of objects, what sort of grasp to use, how to sense the weight of an object, and how to move it so that, if it is full of fluid, nothing spills.

The transition from early navigation systems to complete indoor navigation solutions took less than a decade. Today, software development proceeds ever-faster, and we never face the limitations in calculation that were present even one year ago. If basic manipulation is a solid communal service in five to eight years, the number of skills that we will likely have fully developed within twenty years is staggering. For much of what we humans do in the world mechanically, there may well be recipes for the right robot to imitate.

Primer 5: Connectivity

In California, the Argentine ant, an invasive species introduced unwittingly by human visitors, has become a single, massive supercolony stretching nearly 600 miles along the California coast (Walker 2009). "Californian large," as the colony is known, dwarfs every other ant colony in the state, and presents a unified power that no other ant can combat effectively. Every normal attack fails when your foe has an infinite supply of soldiers and information about food sources—*everywhere*. The Argentine ants all mingle without being attacked—they breathe the same pheromones, and even have redundancy in queens, with eight queens for every one thousand worker ants.

Colonies, whether real like Californian large or science fiction like *Star Trek*'s Borg, are both interesting and unnerving because they have a foundation of power and knowledge that far exceeds what you experience when you interact with just one member. Whether or not the mass of robots that comprise our robot smog will be members of a single megacolony is open to debate. But what is certain is that our robot smog will have massive connectivity with itself and with the information superstructure of the digital world. For all intents and purposes, there will be a new supermassive robot colony, and this presents us somewhat independent-minded humans with awkward future interactions.

The fundamental connectivity argument for robots will stem from the fact that they will be initially mediocre in their capability to perceive and interact with the physical world. To see the world and understand that perceptual signal as well as possible, robots will have to make use of online resources, which will certainly include visual recognition services and specific data-bases of objects and information. Think of this as a giant Google search engine in the service of robots rather than people: *Robo-Google*, organizing the world's information for mobile robots. Robots will want to know how to read traffic lights, how to turn the doorknob of a door they have never encountered, and how to fold or unfold a particular model of baby stroller. They will need to be able to recognize faces, and then register the identity so that they can recall it later.

Once robots collect and store physical-world details using Robo-Google, then there will be a natural progression toward broadly sharing that information among robots. This is interest-ing in that Internet-based information that is not physical will be smoothly associated with the tangible world. A robot sees you struggling with some luggage at the airport and asks if it can contact you about some new power-assist luggage. When you say yes, you will not need to provide contact information. The robot knows your smartphone details—the phone and the robot are on the same network—and your interest in new luggage will be available to any robot or service paying for that privilege. But sharing will not stop there: the market intelligence that you are receptive to polite new product offers will be bought and sold in the robot network as long as it has financial value.

Because of the arbitrarily large nature of online community, when you see a robot on the street it will be hard to surmise the extent to which it has background information about you. One analogy for this situation is that of a movie star or poli-tician meeting fans. The fans know an extraordinary amount about the star, but the star knows essentially nothing about each fan. There is no sense of parity or reciprocity in the relationship;

rather, there is a strangely one-sided conversation that does not feel mutually satisfying.

The final word on connectivity is that robots, even social ones, will be unlike people. To think, therefore, that we will interact with socially intelligent robots the way we do with people is utterly naïve. There is no real precedent for how this will work, and the only prediction I can make with confidence is that the robots you meet on the street in 2035 will know much more about you than you will know about them. If you are an unyielding optimist, you could interpret this to mean that robots will treat you as if you are a movie star.

Primer 6: Control

The stereotypical view of robot control is that there are two distinct modes: robots controlled by people, often called *teleoperated robots*, and robots that are under their own control using local sensors to decide how to move and behave, *autonomous robots*. We are now trending away from both of these extremes and, instead, are heading for approaches that blend autonomous decision making and teleoperated human control fluidly. Some robots will be mostly on autopilot, but will request human help when unsure about what to do. Other robots will be autonomous, but needy—asking passersby for help with directions, doors, and elevators. Other robots will be dedicated teleoperation vessels, under the high level control of humans but avoiding obstacles locally with their own sensors and reactions. Yet other robots will run the gamut from fine-grained remote control to completely autonomous behavior depending on the task at hand.

Starting in 2004, the NASA/Ames Intelligent Robotics Group began a research project called Peer-to-Peer Human–Robot

Interaction. The researchers studied how robots and humans can work together on construction projects on the moon, with an eye toward having humans in the safety of a lunar capsule help robots construct a habitat on the moon's surface (Fong et al. 2005). Sometimes humans in spacesuits would be working shoulder to shoulder with robots to lift heavy parts or inspect a welded joint. At other times a robot might run into difficulty—for instance, bolting together two parts of a prefabricated frame—and it would request remote help. The human operator would need to move from supervisor to teleoperator in such a case, taking over command of the robot's cameras and manipulator arms to directly tighten the problem bolt from afar. Researchers call this ability to seamlessly shift levels of robot control *adjustable autonomy*, and it will be an important design consideration for the robots of the future.

Adjustable autonomy suggests that, as with commercial jetliner autopilots, the interfaces are designed so well that humans can gain situation awareness and take over the controls whenever necessary. Conversely, adjustable autonomy suggests that whenever the task is easy enough for the robot to handle on its own, the human operators can grant it the autonomy to continue without direct control.

In the NASA case, this means ensuring common language between human and robot. "A little to the left" and "above me" would need to be interpreted the right way by both robot and human, so that they can communicate, in English, about a task and about possible solutions to a situation. Teleoperation controls also need to be designed so that sensors stop the remote operator from accidentally doing damage—for instance, failing to notice a ledge while moving the robot and hurting the mechanism. A great deal of research on adjustable autonomy is focusing on safe teleoperation controls, including force feedback and even three-dimensional visualization so that the human operator can be as context-aware as possible.

Adjustable autonomy will be a fact for future robot systems, and this will mean that robots encountered in the home or on the street will have varying levels of human control, from none at all to supervisory to direct remote control. As autonomous robot control and dialogue systems improve, it will be less and less easy to infer whether or not humans are controlling a particular set of robots, and whether the robots, in turn, are being supervised by people or are running loose, in the wild. The control advances of the near future therefore act as a veil in terms of our interaction with robots—the true identity of the robot, as an autonomous machine or human vessel, may be hidden and, with it, our ability to build a concrete model of just whom we are interacting with.

Synthesis

Given all the predictable advances in structure, hardware, electronics, software, connectivity, and control, we can reconsider a vision for the remixed robot inventions of 2035. Anyone who, today, can put together a bicycle from a box of parts or assemble a Swedish bedroom dresser ought to be able to build robots from a multitude of kits. The robots will fly over our heads, hop alongside us, run fast, travel up and down stairs, go anywhere in a home. When you see a robot on a sidewalk, it will not be clear whether it is under remote control by someone nearby, someone miles away, or under the direction of an AI program that is making entirely digital decisions. The robots can make us out, detect our eye gaze direction, and understand most of what we say.

They can take pictures, post them instantly to the world at large, record our voices and sounds, and shift seamlessly from a low-IQ automatic robot to a high-IQ robot vessel controlled by people.

Today's video-recording cell phones are already changing the relationship between citizens and their government. When a policeman fires pepper spray into the faces of a line of demonstrators, everyone sees the brutality online. Nothing is transient, because everything can be captured and shared. Expectations of privacy have to shift because anything done in a public space can be captured and broadcast. As robots decide automatically just what to record and publish, so our sense of Internet identity will have to change. We humans will no longer be the only species that records events and publishes them online. Indeed, not only will robots be authors of new content, they will also be busy consuming Internet content, just like us. When you run into a well-connected robot, it will be unclear from first impressions whether the robot knows everything about your day—where you have walked, how you have interacted with other robots all week—or whether you and the robot are equally uninformed about one another.

These robots could be gaudy, as friendly as an over-eager dog, or completely nonplussed, as disaffected as a Siamese cat. They can be as small as the palm of a hand, flying at five miles per hour and perching on tree branches, or as tall and gangly as a three-meter-high pole that is only centimeters wide. Robots will climb trees, juggle a dozen balls, and wipe down windows three stories up. They will run around in a park, playing catch innocently with children, but looking and roaring like a five-foot-high Tyrannosaurus with six-centimeter rubber teeth and glowing red eyes. The annoyance you feel when a mobile phone rings during a picnic concert may seem laughably insignificant to the ways in which all manner of robots may distract, interrupt, and confuse your plans.

If there is a silver lining in the wild west of custom public robots, it is a hope for the same response we witnessed in the early twenty-first century following the explosion of chaotic, highly animated websites. Design and human–computer interaction were fields filled with vigor in response to badly done websites. These disciplines recruited and trained a diverse set of individuals into the design practices that needed to be taught for effective, interactive websites to come of age. I do hope that human–robot interaction and robot design theory will rise to this occasion, introducing principles and applications that will have a cleansing effect on the robot smog that our creativity will manufacture. But at best this will be reactive rather than proactive, just like government regulation, municipal zoning, and legal jurisprudence. It will likely get much worse before it begins to get better.

3   Dehumanizing Robots

Pittsburgh, Pennsylvania, April 2045

It was the kind of day pilots call "severe clear," with a brilliant, almost California blue sky in Pittsburgh—a rare event—and spring blossoms in the trees all around the park. It was Friday, 4:00 p.m., school had just finished, and it was time to celebrate the fine weather with my son by taking him to the park. Blue Slide Park, only four minutes' walk from home, had the longest slide in the city, built into a hill and painted bright blue. With one hand I was holding the cardboard meant to cushion Jasper's ride and also take him down the slide as fast as humanly possible; with the other hand I was holding my mobile to remind me to call the airline and get my mother's ticket sorted out so she could visit her grandson. I had already carefully explained to Jasper why I was going to be on the phone—since he knew I always hated the sight of parents off in some other space when they were at the park with their own kids.

Jasper tugged on the cardboard and pointed at the street just in front of the park. It was a traffic nightmare, with cars stopped in all directions at the three-way stop in front of the park entrance.

"Papa, look at the blue car."

The blue car—a driverless Honda sedan—had been left by its driver to find a parking spot. Since the gourmet hot dog stand had won a Best of Pittsburgh award, every driver seemed to send their car this way for parking, and so none of our friends could park on the street anymore during dinner hours. The parking spaces would just disappear thanks to ever-patient cars driving around and sending location data to each other about the best spots. Efficient but absurd. This empty, blue car had its hazard lights on—odd for driverless mode—and was doing this strange maneuver. It seemed to be parallel parking in the intersection, edging sideways on the road, and of course totally ignoring the angry honks of the human drivers.

Jasper and I crossed the street and walked closer to view the spectacle. There was some wet, brown paint on the asphalt in front of the car and beside it, and it was maneuvering to go around the wet paint and continue down the road—as if the paint were radioactive.

"Jasper. This is interesting. I read about this last night in bed. You see the wet paint in front of that car? There." Jasper nodded. "It looks wet, like it just rained."

"Right. See how the car won't go near it? Look inside the car. Is there a person driving it or is it a 'bot right now?"

"I don't see people. So it's going to pick people up?"

"Or it just dropped people off and is looking for parking. Or something. Anyway, when people drive a car, they use their eyes and ears and senses to drive. But when the car drives by itself, it uses special sensors that are built into the car. That's how these cars know to stop when we cross the road—they see us using lasers that bounce light off us, and cameras that take pictures of us and analyze the pictures. That wet spot is special paint. It has little tiny pyramids in the paint that absorb certain kinds of light. It's not really wet—it just looks like that to your eyes. And to the car's eyes, it looks like a hole! It looks like the road has a big, huge pothole that would damage its wheels."

"Wow. Doesn't the car see all the other cars driving right over it?"

"Good idea, but these cars try to be supercareful. As soon as one of their sensors says there is a problem, they become very cautious. In fact, see the cars up the hill u-turning? I bet those are driverless, too. They're turning around because this one has updated the map and now all the computers think there is road damage here."

By this time the car had managed to sidestep the biggest hole and was driving on the wrong side of the road, creeping past the hole, and finally pulling out in front of it. Two cars further behind were also driverless, and one had already turned on its hazards to make the same ridiculous maneuver.

"But who put the paint there?"

"That's what I read about last night. Neighborhood groups have been getting angry about these cars coming in just to park. It makes our streets really busy. There's no way you could ever ride a bike here now, Jasper, but five years ago you could mess around on the street in the middle of the day. Anyway so people are upset at the cars. So there's this new form of urban graffiti that I read about that popped up in New York and now it looks as if it's here! Some artists posted online how to make this paint. Now you can paint a hole anywhere. It changes the local map, and in about an hour there is much less traffic on the road. So it's all about cutting down local traffic, but of course it's bad."

"Why is it bad? Did you help them?"

By this time the attraction of the park outweighed the fun of watching another car act silly, and we were approaching the street to cross the road and enter the park.

"No! No. I didn't. It's bad because the computer updates also make the road crew come here, and they come and realize there's nothing to repair. They have to burn off this paint, or paint on top of it, and then they look for the vandals because they want to charge them for the repair—it's expensive for the crew to come out here."

At the road I reached out to grab Jasper's hand and realized the phone was still there. Dunce, I already forgot I meant to call the airlines. "Jasper, I'm going to make the phone call now for Grandma. Listen, just stay in that part of the playground and I'll watch you from here while I'm on the phone."

Jasper ran ahead to the climbing wall as I phoned customer service.

"This is Blue Sky customer service, where we always give you personal attention. Hi, I'm Ruby. Would you give me your name and let me know how we may help you?" This personal attention nonsense was annoying. Everybody says that. Just what do they define as 'personal' now?

"John Nack. John. J-O-H-N. Nack. N-A-C-K. Complex seating arrangement for elderly person. Requires specific type of aisle seat. Travel times flexible. Speak to human."

Jasper had climbed off the wall. I scanned the area. There were maybe a dozen kids standing on a grass berm, looking north into the park. Jasper had joined them, to the left, staring at something. I started walking up the hill to investigate.

"My name is Ruby. I can help you with that, Mr. Nack. May I call you Mr. Nack?"

"No. Negative. Request human. Now. Operator." In the old days you could just press 9 four or five times and eventually you'd get to a human. Not anymore.

"Mr. Nack, what are the departure and arrival cities?"

It looked like a football game at first, but that wasn't it. Four or five teenagers were laughing, standing around the remains of an old monkey bar set. One of them whooped and threw a stick off to the left, and then I saw a four-legged robot run hard after the stick. They were counting down, shouting "Three, two one!," and right as they reached one the robot froze in midair, as if stopped by some force field, and fell to the ground on its back.

It twitched for a few seconds then righted itself and scanned around, probably for the stick. I edged closer and caught up with Jasper.

"Oh, come on. Okay, fine. San Diego to Pittsburgh. She has a bad right knee that cannot bend. I need a left aisle seat near the front of the airplane. Any departure in early August is fine. Two weeks' total stay. Cheapest possible ticket."

"Mr. Nack, may I interest you in a first-class ticket? We have excellent aisle seats available in first class."

The robot dog walked back to the old jungle gym, and I noticed the rope. They had tied some sort of rope to one of its rear legs, and it looked as if it didn't have a clue. When it got back to the gym, the kids brought the stick back and did it again, this time throwing the stick up into the air.

The robot tracked the stick, ran, and jumped, but as soon as it was airborne there was a sickening cracking sound and it spun and hit the ground with enough force to bounce—and it was easily a fifty-pound 'bot. This time it took longer to get up, and something was wrong with its rear leg—it had a strange, bouncing gait.

"Hello, Mr. Nack? Which class would you like me to search?"

Great, now it's upselling to first class. This is stupid. "Help. Human. Manager. Kill the damn robot. Give me an actual human being. Help."

The kids threw the stick straight out again, this time really far. The dog managed to accelerate to what looked like 10 miles an hour at least, and the rope snapped tight. This time the rope came off and the 'bot tumbled. But as it got up it was obvious that the rope had not come off; the leg had actually ripped off its torso.

"Human help. Complex purchase. Urgent. Now."

"Mr. Nack, I am a human agent. Let me help you. We do have a special offer on first-class seats. Please, let me help you. I can also look at aisles in the back."

"Horseshit! You're as human as the robot that fixes my car. You're too stupid to be human."

I swear I heard a gasp on the phone—an audible gasp. I hung up immediately—it was an automatic response. My heart was pounding, and I felt sick. She was human, and I'm an idiot. That was really stupid.

Out went the stick again, with cheering from the kids and a big countdown. "Jasper, turn around. These people . . . they're sick. Let's go home."

• • • •

Human–robot interaction researchers study relationships between robots and humans, often by running psychological experiments in which they observe human reactions to unconventional robot behaviors. One present-day study measures empathy and moral standing between adults and robots. Children are introduced to an apparently autonomous mobile robot, and together they play a children's game called I Spy. In the midst of game-playing, lab technicians come in and tell the person and robot that it's time for the robot to go to the closet. The robot complains, saying the timing is unfair and that they are in the midst of a nice game. The technician is firm, and in spite of constant complaints pushes the robot into a closet, turns it off, and shuts the door as the robot ineffectually says, "I'm scared of being in the closet. It's dark in there, and I'll be all by myself. Please don't put me in the closet" (Kahn et al. 2008).

In another human–robot interaction study, researchers wanted to measure the level of destructiveness people can unleash upon seemingly intelligent robots. Student volunteers are introduced to a toy robot that follows the beam of a flashlight and are encouraged to spend some time playing with it. They are told that their job is to test the robot to verify that its genes are worth replicating. The researcher lets the student play with the robot for some time, then announces that this robot is substandard and must be destroyed, giving the student a hammer and literally asking him to "kill the robot now." Researchers then measure the level of destructiveness by counting the number of total hammer hits and the final number of fragments of crushed robot (Bartneck et al. 2007).

These are early, and very eerie, projects dedicated to understanding how we think of robots, and where we place seemingly autonomous robots in our system of ethics, empathy, and action. In science fiction, the problem is reversed. In Philip K. Dick's novel *Do Androids Dream of Electric Sheep?* (1968) and the movie adaptation of his work, *Blade Runner* (1982), bounty hunters try to find and destroy renegade replicants—engineered beings with android brains. But replicant engineering has advanced to the point that these creatures are nearly indistinguishable from humans, yet they are enslaved in a system whose moral justifications are failing. Dick invented the *Voigt-Kampff* machine as the key tool for discriminating humans from replicants in this world. The machine works by detecting physiological responses to carefully worded questions during an interrogation, measuring the empathic response of the subject. So in fiction, the one solitary gap that remains between humans and androids is emotional: empathy. Of course, the story is powerful because it breaks down even this one final discriminant, leaving us to question the extent of our human rights and liberties.

But the irony is that, in our present-day nonfiction world, researchers are still busy trying to ascertain our human emotional response to robots. We do not even understand human empathy in the mixed-species world of humans and robots yet, let alone the emotional qualities of robots themselves. What makes this form of robo-ignorance even worse is that we do not fast-forward to the *Blade Runner* world overnight—rather, we will spend decades in intermediate stages, where the early robots out of the "womb" will be inferior to people in numerous ways, yet they will be social, interactive, and incorporated throughout society because they are useful enough to turn a profit for someone.

How will we treat these pioneering robots, which will doubtless have characteristics we can easily take advantage of if we so choose? We can extrapolate from examples of truly autonomous robots that have been introduced to the public in the past decade.

One experience that has always remained with me involves my undergraduate research robot, Vagabond, exploring the sandstone arcades of the central quadrangle at Stanford University. Our goal was to create a navigation program that would enable Vagabond to travel anywhere in the quad, and we had gone so far as to measure and map, by hand, the complete layout of the quad—the position of every hallway, curb, and pillar down to the nearest centimeter. Navigation software enabled Vagabond to measure distances to walls and columns using sonar, estimate its position in the handmade map, then navigate the walkways to a goal destination. Along the way, Vagabond used the same sonar sensors to detect people in its path, stopping or patiently attempting to navigate around them while continuing to track its position.

Normally, we communicated with Vagabond from a desktop computer on a wheeled cart. We had very long extension cords connecting the cart to our offices, on the second floor, and were nearly always in the same long walkway as the robot. Many tourists would walk by this picturesque location, and so we became very good at compactly describing our research dozens of times a day.

On one occasion, we were pushing Vagabond's navigation to the limit, having told it of a destination several buildings away. Our cart was situated near its starting point, and the robot was already out of sight. After a chat with Ben Dugan, my colleague on the project, I decided to check on the robot and went around the corner. I saw the robot—a two-foot-high black cylinder with a Powerbook 150 laptop latched on top—at the far end of the corridor, with two people standing next to it: a tall man in cowboy boots and a woman. I was 25 meters away and as I walked toward Vagabond I realized what they were doing. The woman was blocking the robot's path, keeping it still, and the man was kicking the robot on the side, hard. Hard enough that the robot was tipping and righting itself on every kick.

I started running, and as I neared them they began walking away and the man said, "I'm still smarter." In all our programming, in all our obstacle-detection logic built into the LISP code, we had never accounted for this particular possibility—man kicking robot to show off to girlfriend. It was a turning point in my realization that robots will cause human behavior that we may find very surprising indeed.

My second personal experience stems from Chips, an autonomous tour-guide robot our research group installed at the Carnegie Museum of Natural History (Nourbakhsh et al. 1999). Chips provided multimedia tours of Dinosaur Hall from 1998 to 2003, playing videos of paleontologists and dig sites while also traveling around the massive Tyrannosaurus type specimen at the museum, pointing out details concerning the bones and additional exhibits on the walls. Chips was tall and heavy—more than 2 meters high and 300 pounds, so it was critical that this robot play it safe in a space full of children and strollers. Its navigation system was designed so that any obstacle in its path brought it to an immediate, full stop as it said "Excuse me" through its speakers.

But during the first few months of deployment, we found Chips facing the same pathological condition time and time again. Children would be following the robot as if it were a pied piper, attending to its videos and enjoying the spectacle of a massive mobile robot with a cartoon face. Adults would step in front of the robot, watch it suddenly halt and say "Excuse me," and then wait there, smiling. And wait. And wait. Those following the robot on tour would eventually be fed up with the delays and leave for greener pastures.

Once again we had been naïve, assuming that "excuse me" would mean, to people, "Please step out of my way." Experimentally, when a robot tells a human "Excuse me," the person often interprets the statement to mean "Hey human, look at you, you have the power to stop me. How cool is that! Play with me."

The solution to Chips's abuse problem, obvious in hindsight, was a simple phrasing change from "Excuse me" to "Excuse us. You're blocking my path, and I am giving a tour to the people behind me. Please let us continue." What a difference. People would block the robot, listen to its response, look at the people behind the robot embarrassingly, and move right out of the way.

I never really discovered a way to make people treat the robot with more respect. I simply brought the people following the robot into the social equation, and manipulated the human obstacle into behaving more politely for the sake of their human cousins.

So there is a chance that even slow robots will be treated well by people when they are wrapped into a human social context.

But the story may be woefully different when robots are out and about on their own, apparently autonomous and disconnected from the social fabric of real people.

The legal code would be a particularly bad place to look for hints regarding people's interactions with autonomous robots. Last year, following encouragement from Google, the State of Nevada enacted legislation to soon make it legal for autonomous vehicles to drive on the state's highway system (State of Nevada 2011). To date Google's autonomous driving machines have already covered more than 200,000 miles in California, where there are no laws explicitly forbidding robotically driven vehicles. And yet the diversity of ways in which legal boundaries, human behavior, and robot cars on the road will intersect are not predicted by Nevada's legislation or by existing bodies of law. In August 2011 the automotive blog *jalopnik* broke the news of a fender bender caused by one of Google's autonomous cars. Google issued a statement explaining that the accident was caused by the human in the Google car, since he was driving manually at the time. But, of course, this response begs the question, how will blame be assigned when the robotic driver and human are both partially responsible? Toyota's Prius brake issues in 2010 and the infamous Audi 5000 reverse-gear episode in the 1980s demonstrated that complex machines can interact with people to cause complex failures. In each case, the companies initially blamed people wholly, then eventually reengineered their machines to make them safer. Even when people and intelligent machines work together, failures will sometimes happen and responsibility for damage will be murky.

But the vignette in this chapter suggests something more nefarious when humans and machines are not on the same team: creative forms of robot abuse. Social barriers against such abuse will be weak, even when robots are others' property and the abuse is a form of property damage. We can be quite sure that early robots, whether cars or park toys, will be easy to fool using crude human tricks, and there will be a plethora of willing people interested in testing legally indistinct boundaries to entertain themselves at a robot's expense. The gates will be lifted for the early release of robots into the wild, and any laws governing human–robot interactions will certainly be born as reactions to unexpected side effects rather than as proactive laws designed in advance. Law will not guide good human–robot relations.

If not law, how about ethics then? Last year I taught a new Ethics and Robotics course and invited Professor John Hooker, a business professor specializing in the formalization of contemporary business ethics and the analysis of historical business practices, to be a guest speaker. When asked about the ethical requirements governing human–robot interaction, he drew upon a historical analysis of slavery, noting that the ascription of *agency* to the slave made it unethical to treat the slave as anything less than a human being with full rights and privileges (Hooker 2010).

The concept of agency does indeed have strong roots in historical debates of human rights; and it is also a term that is frequently used in design and robotics to refer to made artifacts that are beginning to show signs of decision-making competence. The concept of *human agency* formally rests on two requirements: the capacity to make choices, and the power to enact those choices in the world. Agency is decision making followed by action. In the Kantian will theory of human rights, people must have an inalienable right to exert their control—to exercise their will by making decisions and having the right to follow through with action. By this metric, the slave is dehumanized in one of the most unjust possible ways, because agency has been withdrawn from that person and, with it, the power of the slave to be responsible for and authoritative over his or her own actions.

Extending this concept of agency to robotics, Hooker suggested in my class that as soon as people ascribe agency to a robot, then from an ethical point of view people must provide that robot with the same rights as humans. His analysis is interesting in that it privileges peoples' views of agency far above the nature of how robots actually work. Professor Hooker's point is not that robots equal humans, but that if we begin ascribing agency to robots, and treat those robots unjustly, then we are unethical, and we will be inconsistent with our own moral charter.

Although this is a provocative way of thinking about ethical human behavior in the face of artificial intelligence, the scenario is many decades away at the very least. The boundary case, once again, is where we find this analysis problematic in the near future. Early autonomous robots will show hints of agency— they will have goals and will endeavor to achieve those goals. Importantly, they will enact those goals in the real world, and this is what sets these robots apart from computer-based software agents. But these early, embedded robots will also show signs of being significantly inferior to humans in many ways, from their ability to perceive the world to their ability to learn and adapt. Will we treat these robots as if they have human-level agency? Certainly not—they will be executing tasks for us, and we will treat them as personal property.

The complication, then, arises from the collision of the ethical and realistic analyses. We will treat these creatures as personal property, and yet over time they will show more agency and more humanlike disposition than any other creatures we have ever built or tamed. The question is, will our treatment of robots change our ethical balance—will our principles be subverted by our very own desire for the smartest robotic slaves money can buy? And will this impact how we treat other human beings? Is there a danger that high-functioning social interaction with robots, whom we cerebrally view as totally different from people, will derail the way in which we have high-functioning social interaction with people?

Some will argue no. Parents can spend endless days talking with their three-year old, yet are capable of jumping right into adult conversation, happy to finally have a chance to speak with a mature person. Pet owners can establish meaningful, deep relationships with dogs, yet this cross-species relationship does not negatively color interactions with humans. In *Alone Together*, Sherry Turkle argues that our deeply emotional relationships with robot toys and our addiction to superficial but broad online networking have a profound impact on our self-identity and the depth of our real-world relationships (Turkle 2011). How we interact with each other, face to face, is already changing thanks to new forms of techno-interaction that pervade daily life.

But what happens in two decades, when we are conducting many of our regular affairs through robot proxies? We will purchase airplane tickets in conversation with robots. We will order dinner. We will argue about a missing utility bill or a problem with the car that the robo-mechanic failed to fix. The machines are about to graduate to a new plane of relationships with us, and we are still busy understanding how a toy robot can twist us.

These next-generation robots will be inherently schizophrenic—sometimes autonomous, sometimes telepresent and under the direct control of real people. With all manner of robots, we will find ourselves entering the same negotiations and arguments that we also do with people, although we will recognize full well that we are dealing with subhumans. This will be new territory in testing our ability, as humans, to flexibly change interaction modalities as we switch the species we are talking to. The danger we face is in how mediocre interactions with robots poison our social habits more broadly.

The history of war and careful laboratory psychology studies have already demonstrated that as we dehumanize other peoples, we remove our own barriers to acting with cruelty, torturing, and harming. But never before have we had the chance to dehumanize socially agile nonhumans. If this happens, then we face the odd technological side effect that, as we dehumanize our relationships with robots using a rather broad brush, so we may incidentally dehumanize our relationships with people.

4 Attention Dilution Disorder

Washington, D.C.; San Francisco, California; Tuileries Gardens, Paris; Cheddar Gorge, U.K., September 2050

"I just think it's stupid that they need you there in person. I mean literally."

"Honey, it's a deposition. They're always in person. Besides, we make great money on the consulting fee. Every hour pays for, what, probably ten of these 'bot trips."

"I just wish you and I could be walking in person at Reservoir Park and you could deal with them through a 'bot."

"Yeah, but we'd still want an hour in Paris, right Katy? So this is the life."

"So what's the story tomorrow? When do you get back?"

Paul had been listening to his wife in stereo mode, with both ears. He likes to avoid the bi-aural mode, where you have two ears on two different conversations at the same time. His agent knows this and notifies Paul with a vibration in his right pocket that it needs help. His video glasses flash a message in the corner of his retina— *Need Help: Julia.*

"Paul, I said, when are you getting back here to the Bay Area?"

The agent flashes a green light in Paul's eyes to confirm that it can deal with this question autonomously, and Paul twitches his right thumb to select Auto-Complete. His small gestures in midair are picked up through his sleeve as if he were typing on a portable keyboard. He hears the agent start to answer Katy in his own, digitally reconstructed voice, *Let me look the flight times up. Hmm . . .* satisfied, he spins his right hand and selects the Cheddar Gorge video and audio. The video is already locked in on Julia's face as she speaks.

"Right, Paul? I mean, he's an investment banker. Even you're more fun than he is. Wait, that came out wrong."

"Hah. Are you genuinely happy when you're with him, Julia? You're a smart, beautiful girl that climbs cliffs all over the world—in person. Don't just settle."

"Boy, that's pretty clichéd. Am I talking to your agent? Kidding. Ooh, watch this one. Do you have that thing dialed to human performance? Don't cheat!"

Paul scans the rock face by moving his head; the camera copies his motions remotely. He has the 'bot set at 5.10c. Fair enough. He clicks on two handholds with gestures in the air, selects Auto-Climb on the Special Actions menu in the corner of his left eye to turn the rock climbing 'bot loose, then swivels the camera up to take a look at Julia. Incredible shape she's in. Julia is above him to the right, he can see her jamming for the next move and dart a look down at his 'bot's camera.

"Hey, look at the rock face! Don't cheat and see what handholds I use. No fair."

"5.10c. Honest, Julia. Man, I am sorry I couldn't be here in person after all. Sucks."

"You'll make it up to me next time, right?"

"Of course. But you know Katy. She gets jealous about everything."

"Even a 'bot visit? Come on, I've known you longer than she has."

Another notification vibrates and pops up on his right eye—*Need help with Katy in Paris*. He selects Auto-Complete with his fingers and spins his hand back to Paris. The Paris video returns to full size, but before switching over with a Confirm gesture he takes his leave.

"Hey, Julia, I need to take a call—about the deposition."

Paul doesn't wait for an answer—no time—and hits Confirm to flash fully over to Paris.

"Well, I can't wait. Think about it. Oh, and look how cute he is. Dimples!"

The 'bot must have understood Katy's words well enough to realize it ought to be co-gazing. It is looking at the same spot as Katy's 'bot, watching a toddler, stark naked, who is walking in a shallow fountain, laughing.

"Beautiful. Cute. Katy, I really love walking here. Keeps me sane."

"Well, I have a special sanity recipe just for you. At 6:00 p.m. tomorrow, like you said."

Another vibration. Agent needs help on Julia.

"Honey, I'm dry. You mind if I find a coffee here? There are, like, ten coffee shops per block."

"Sure, go ahead. I'm sipping a mint tea in our backyard. Blue sky here. I'll send you a snapshot."

Paul selects Walk-Alongside and Auto-Complete in special actions, then flashes out of Paris with a hand spin. Instead of selecting Cheddar Gorge, he turns up transparency from 50 percent to 100 percent on his left eye and looks around.

There really are tons of cafes there, all variations on the same Portland-comes-to-D.C. theme. He steps in one and into line, flashes over to Julia rock-climbing, and tries to speed-read the agent conversation summary that has been on his left eye for the past two minutes.

"What was that, Julia?"

"I'm just saying, it's proof that we can be just friends."

"No, you're totally right. I just think Katy probably had different experiences in her college years. Maybe she never had a totally normal male friend."

Paul swivels the camera up again. Julia looks even better than last summer. He snaps some pictures for his library.

"There was this one guy. Kurt—did I tell you about him? Freshman dorm, we all thought he was the most annoying guy in the building. And here's the thing, he was an amazing rock climber, so we spent hours at the indoor gym belaying each other. The second you walked into the gym, he was nice—focused on climbing, you know. Like a professional. But every time you saw him in the dorm or anywhere else, he was unreal. Full-scale annoying. Loud, with a totally grating laugh. So I finally thought . . ."

Paul hits Auto-Complete again and concentrates on the transparent view in his left eye—the café interior as he reaches the front of the line.

"Hi. A dry single cappuccino and that almond biscotti please. For here, thanks."

He wanders out into the courtyard and sits at an empty table to wait for the drink, stretching his neck. Julia's into a long story, so he switches over to Katy.

"Hey, what do you say we sit at the bench by the roses? Wasn't that where you told me about your flying issues for the first time?"

"Oh, come on. Sure, let's sit and watch. Do you remember this many ridiculous dogs before?"

"Oh yeah. Even in restaurants, remember?"

Paul selects the destination bench and sends the two 'bots ambling there slowly.

"Paul, honey, my mom's calling. She's got it on Very Urgent—give me a minute."

Paul looks around the café and mutes Paris. Just next to him, there is a girl with no heads-up glasses, no earplants. She has a character he doesn't recognize tattooed on the small of her neck. With a pencil, she is sketching a leaf she has pinned to a board on her table—maybe a botany student?

"Excuse me. Hi. Can I ask you a question? That's a figure I don't recognize. Kanji?"

She smiles—he could see this from behind, just from the movement of her cheekbone—and shakes her head.

"Have you heard of symbox? This is the character I designed for it."

"Very cool. I read something in the *New Yorker*—this is a classification for species, right?"

"No, it's a description for phenotypes. You know biology? I proposed this for Rhesus monkeys and it was accepted."

"And, but, you're a botanist too—I'm guessing. Sorry—I don't see any leaf sketchers at the other tables . . ."

Katy takes another sip of mint tea and tries to reason with her mother again.

"All I am saying is, you can't mark the message as urgent if it's not urgent. How will I know if there's a real emergency? It's like the boy who cried wolf, mom."

"Kathryn, if I don't, you will just make me leave a message or talk to your agent. Honey, I really want you and Paul to come over for Thanksgiving. Just say yes and we'll schedule it all with your agents."

"Mom, of course we'll be there—just let us use 'bots. The trip is such a waste of time on an airplane."

"It's not the same, Kathryn. We have nice conversations with you in 'bots, but I want you to taste the dinner. It's been a full year."

"I'll talk to Paul—but, please, it is just less stress all around if we can 'bot in and sleep in our own bed and have some time together, too. Mom, the time we chat and spend together is what really counts. Please understand. Hey Mom, they even have some 'bots now that can sit at the dinner table and eat with you—they can do everything we would do at the table—we can even do a toast through them."

"That is the dumbest thing I've ever heard, Kathryn. Why don't you just send some random person to my house to eat my food then? What good is that?"

"Mom, we would be at the dinner table participating completely. It really feels pretty good— I tried it out last week for lunch with a friend in Jersey Island."

Meanwhile, Paul gets a notification that Julia's monologue has ended. He flashes to Cheddar Gorge with video and audio and skips reading the story summary the agent has prepared for him.

"So have you heard from him recently?"

"You mean since bumping into him last week? Duh—no. I took two days coming here from Chile, and if I saw him here, too, that would mean he's actually stalking me."

"Sorry—I meant—never mind. Oh, wow. So what do you think of the view?"

Paul's 'bot and Julia are sitting at the top of the Gorge, surveying the landscape.

"Doesn't this give you perspective? I get calmer every time I am up high. Just like flying an airplane. You can't fall deeply into your problems when you're above them. Okay, so he's the nicest investment banker I've ever met. Is there something other than 'yes' or 'no' that I can say to the proposal? Are there other options? Teach me, teach me! You've avoided this with Katy for, what, three years?"

Agent notification—Katy needs an answer the agent cannot construct by itself. Tattoo Girl has turned around and is studying Paul intently, watching his neck movements and his finger movements.

He must look half-crazy poking at the air. He switches back to hear Katy.

"Again. Every time she does urgent. It's just not safe—for her, I mean. Right? Paul. Come on, am I talking to an agent? Don't do that to me. Paul. This is important."

Tattoo Girl speaks with perfectly bad timing, "You're pretty good with that thing. Are you on two 'bots at once? I've never seen that before."

Paul smirks toward her, then answers Katy. "Is it the visit-us-for-real trope again?"

He hits mute on the Microphone menu, looks at Tattoo Girl with his left eye.

"Two 'bots and one call. I'm stuck here and I'd rather be rock climbing."

He hears Katy's answer in his right ear while he's talking, "Yes, of course Paul. November."

"So are you, like, a businessman? Does it ever quiet down? Are you ever in one place at a time?"

He makes an exasperated face, shrugging his shoulders while unmuting Paris so he can respond to Katy.

"Did you tell her I have work? We can spend 'bot time with them and do our work and it's the best way to schedule a visit. And you don't have to fly that way. When is she talking about?"

"Thanksgiving. I said November, Paul. Pay attention."

"Fine. Just like last year—that worked perfectly. We'll send them a nice set of flowers for their table."

"Paul, once in a while, we really ought to see them in person."

Frustrated, he switches over to Julia.

"Julia, you're a beautiful girl full of life and, hey, look, on top of the world right now, literally. Just take it with a laugh—tell him you love hanging out with him and you have many more adventures in you. Makes you seem all mysterious and also into him enough that he doesn't walk away."

"But do I want to lead him on? Or cut the cord? I can't figure it out."

While Julia is talking, he thinks of a good solution for Katy. He switches back to France.

"Look Katy—I can schedule a meeting that morning so it's really clear-cut. We don't have time to fly, but we do have time to chat during Thanksgiving dinner. Perfect?"

Tattoo Girl is still staring. He smiles back, turns off the video entirely, and mutes the microphone but turns on all audio lines with the Mix All option.

"I still have to eat somewhere real tonight. Live. Have you ever used one of these?"

"Never—the rental plans are crazy."

He hears Katy in his right ear. "Okay—go ahead and plan a meeting, Mr. Fixit. I'll tell her you have one. Small fib."

"I have a proposal for you—there's this restaurant that makes a hybrid experience out of it. You have plans tonight? I'm sorry—I haven't asked you your name."

"Malik. And you?"

"Paul."

Now it's Julia's turn. "You know, you're almost helpful Paul. But not quite. Hey, I have an idea, why don't you meet Matt and tell me what you think? We can have a drink tonight together—the three of us."

"Sorry, Julia, I have plans tonight." Paul glances back at Malik.

"How about it—they feed us and we 'bot walk through a local garden in the country where the recipe originates. It's awesome, and we can use my account. You free tonight?"

"Hey, I thought you're in D.C. Does Katy know you have plans tonight? Tell me the juicy bits!"

"Oh come on, Julia. It's business. It's always business."

"Okay—sure. But I don't think I'll like the robot bit. I'm sort of a Luddite. Look, I'm using a real pencil. I bet you don't even own one."

"Julia. Shall we downclimb?"

"No—let's rappel down."

"Okay—you first. Ladies first. Ladies first."

"Shall we walk back to the front gate, honey? I think our hour's up soon. What are you doing next?"

"Preparing. Preparing for the interview tomorrow. Got to make sure my answers are all consistent."

Paul clicks Auto-Complete on both agents, gives the 'bot in Paris the garden gate for navigation endpoint, and puts the 'bot in Cheddar in Autonomous-Rappel mode.

"Malik. It'll be a new adventure. I am a Luddite too, in my own way."

Paul's cappuccino arrives. Perfect. It's dark, almost chocolate-colored on top. He takes a sip, leans forward, and smiles at Malik, who has turned her chair and is facing him now, pencil still in hand. "So."

• • • •

Humans have always been jugglers. We undertake as many physical activities as we can, simultaneously: we walk, chew gum, hold hands, and fish for change to purchase ice cream—all at the same time. Cognitively, we juggle just as much: looking over the restaurant menu to decide on dinner, chatting with our friend about the couple at the next table, and glancing at the smartphone on the table to see who emailed the latest urgent request.

Robotics will change the way we can juggle dramatically, both because it changes our relationship to communication technology and because it extends our physical reach beyond where our body happens to be located. There is a fertile subject of research in robotics today that studies exactly how to extend human reach both cognitively and physically, and this is called urban search and rescue (USAR) robotics.

USAR begins with the premise that disasters will never go away—whether they are natural or man-made. The immediate aftermath is largely the same: people are trapped in dangerously unstable disaster zones, and every minute counts in the attempt to recover victims before hypothermia and injuries result in fatalities. The disaster sites are no longer accurately mapped—any blueprints of standing buildings are suddenly and horribly out of date. Moreover, the air and the chemical status of the site are in question—there may be lack of oxygen, acids, or flammable liquids that make entering the zone too dangerous for emergency crews. The remaining crawl spaces and passageways may be far too narrow and unstable for rescuers—in fact, walking on them may cause further collapses that kill victims trapped below.

All these conditions make USAR a compelling candidate for custom-made rescue robots that would be deployed immediately into the accident site to find pathways, map the structure, and find victims. The robots could contact the victims, letting them know that help is on the way, provide them with water and food, and even set up direct communication links to rescuers on the outside so they can assess medical conditions and target the neediest first for invasive rescue operations.

These needs also demand sophisticated robot capabilities, and many research groups are hard at work trying to deliver robots suitable for USAR operations (Linder et al. 2010). Some robots are tracked like tanks; others are legged so that they can climb rubble and even push off narrow cracks. Yet other USAR prototypes are long, actuated snakes, with lights, cameras, and environmental sensors in the head that enable detection of humans not only visually but also by heat signatures and even carbon dioxide emissions from breathing.

In parallel with the strong research efforts on robot hardware and electronics for USAR, there is an equally important research strand undertaken by many robotics and human factors experts in the regime of human–robot interaction. Rescue workers have immense prior experience and skill in determining the status of a disaster site and handling first contact with a trapped victim. The interaction experts use simulators and even physical, manu-factured rubble structures to design and test how human rescuers outside the disaster could control and communicate with USAR robots inside the structures (Lewis, Carpin, and Balakirsky 2009). Part of this human-factors research has to do with how humans can "log into" robots. How can a human quickly gain the situ-ation awareness to see through the robot's eyes and assess the circumstances, given that the operator does not have the entire thread of experience of having traveled to that destination in a first-person way? Panoramic visualizations, high-fidelity imag-ery beamed into heads-up displays, and additional sensor infor-mation such as temperature, slope, and ambient noise are all being used to make the operator feel present enough to make good decisions (Lewis and Sycara 2011).

This research is interesting because the relationship between robot and human is complex. The robot does not move like a human—that is its whole advantage. Therefore direct remote control of the robot by the human is very inefficient—we are not snakes and are quite poor at manipulating a snake's spine. Yet humans may have important intuition about which way the snake should go—where in the structure there are likely to be trapped survivors. So in the ideal world the human may, at times, provide general directions to the robot, but the robot ought to wriggle through the rubble mostly autonomously.

Early simulated tests in USAR scenarios showed that the rescuer in charge of a single robot spends most of her time waiting for the robot to get to the target location, and this is a waste of time that no one can afford in real disaster conditions. The obvious improvement, which is where such research now concentrates, is to equip each human rescuer with a whole team of robots, all of whom are to be managed as they semiautonomously explore and search in the disaster site. Now the human can achieve far more productivity than was possible before, effectively acting like a many-tentacled creature that is reaching into the disaster scene thanks to a team of robot proxies.

In USAR, the effective number of robots controlled by a single human operator has a formal term: *fan out* (Steinfeld et al. 2006). Ironically, fielded robots have very low *fan out* scores today. For instance, the Predator-class drones, unmanned aerial vehicles that fight proxy battles for the United States in distant lands, have a fan out of less than 0.2. That is, more than five people are required at all times, just to manage a single robot. In USAR, researchers have begun to demonstrate ever-increasing fan out—exceeding 6.0—by providing the robots with more and more autonomy so that the human operator is only responsible for the most strategic decisions, with robots making every tactical choice. Critical to this success is the ability of robots to decide when they need to ask for human help—when they face a survivor, or are stuck in the rubble in a way that the robot cannot extract itself, or when the robot has suffered a serious hardware or software error. This "intelligent reasoning" for deciding when to ask for help means that one human can manage even more robots to achieve a higher fan out, even if these robots do, at times, need help. They do not need true autonomy so much as a willingness to call for help whenever required. This alleviates the pressure to create perfect robots, and instead good-enough robots can play meaningful roles in a USAR team because humans will bridge the gap between the robot's capabilities and what the situation demands.

USAR robotics research is a relatively young field, having founded a regular national contest series only a decade ago. In that time it has witnessed significant advances both in the physical creation of robots that, one day, can meaningfully explore disaster sites and in terms of human–robot adjustable autonomy interfaces that achieve high rates of fan out together with the very fast, effective assessment of a structure and its inhabitants. All this extends our ability to explore many spaces at once with the help of robots.

Of course, we are not strangers to technology aids for juggling multiple tasks. In computer science, the quick switch between different thought processes or modes is called a *context switch*. In computer architecture, context switches are key to how we can ask our computers to do many things at once—copying files while checking for new email while also ensuring that the just-inserted USB memory stick is virus-free. Communication technology provides us with a never-ending march toward more and faster context switches because each communication act becomes smaller, shallower, and faster to process. A letter in longhand takes minutes to write, days to wait for, and minutes to read, perhaps relaxing with a tea. The telephone upped the ante to multiple conversations with diverse people in the space of a half hour. You can have three very different conversations that transition you between three unrelated emotional states. Email reduced the context switch time to perhaps twenty seconds—power email users effectively dispatch hundreds of emails a day. Instant messaging, SMS, and tweeting all further sharpen human context switching because each message requires only seconds of investment.

All these technologies have increased our apparent productivity, and often the unrecognized cost is that each communication act is more superficial, less worth the time we spend on writing, reading, and responding to it. Nobel laureate Herb Simon articulated the concept that attention, not information, is now our scant resource. Information and communication are now plentiful, so much so that the challenge we face personally lies in deciding what merits our limited amount of attention. The challenge each corporation faces is in convincing us to attend to the information it values for economic gain. But we have always had another scant resource, built into the physics of our arm's-length reach: we have a physically limited impact on the world. We cannot manipulate the far-away world to hug Grandma, salt an icy driveway, or tuck in children and turn down the lights.

What USAR-style advances in robotics bring to the table is the concept of physical fan out: what if people context-switched, not only to deal with text messages, but also to be in different places at once. What's more, what if robots provided some measure of autonomy, so that you could increase your personal fan out well beyond 1.0. What if your life expectancy was not measured by years to death, but by the extraordinary ability to lead as many lives as possible, so that forty adult years, with a fan out of 5.0, resulted in two hundred years' worth of adult life experience? Of course, physical fan out, like information, will further overload us. Attention will be scant, due to the overabundance not only of information but also of the opportunities to physically act elsewhere—anywhere.

CEO of Me, Incorporated

By 2050, robotic machines with significantly human physical capability should be available. Robots will be able to go everywhere people go, and they will be able to manipulate objects with at least the dexterity of the human hand. We will be approaching the point at which, from a mechanical point of view, robots can extend a human's physical presence with high fidelity. Yet that physical extension will be of limited benefit if a human's undivided attention were required to operate that robot in real time. But such direct control will be unnecessary. By 2050 the perceptual and cognitive intelligence available to any tangible device will have advanced just as much. Robots will be able to take care of the motor control details of common activities, such as walking, running, and manipulating the objects in a home. Cognitively, dialogue systems will be able to track and label conversations between people, and engage in directed conversations using human language. Visual perception will have advanced to the point that identification of objects throughout the natural and designed worlds is a solved problem.

Human interface systems, thanks to the advances of communications technology, will be in spaces hard to imagine today. It is enough to note that immersing oneself in audio, visual, and perhaps tactile realities thousands of miles away will be de facto aspects of how we plug into each other's lives for social and business visits. If robots are beginning to bridge the real world and the Internet world, then the physically realized, home-capable robot will begin to bridge our own physicality of location. *Being-there* and *not-being-there* will become a blurred distinction; just where we are at any given point in time will have less meaning than ever before in our cultural experience. The excitement and anxiety of this new mode of physicality is wrapped up in the reach we will have. Unlike Skype, Face Time, and all other communication portals that place an image and audio source on the table or in a pocket, this time the space traveler will jump off the table, out of the pocket, and push back on the distant world with his own volition, sharing physical space with locals more literally.

Three key ingredients combine to offer this *blurred physicality* in which people may interact with the world: the immersive human interface system, the spatial extension provided by physical robots, and the seamless adjustable autonomy offered on those robots. Together, these three suggest a future in which people do not just context-switch, but live multiple lives simultaneously, thanks to the ability to deputize robots that act out a portion of each life for them. The robot takes a walk with my friend and reports back when it is time for me to join in. The robot AI agent finishes a conversation for me because the key bits are done and now the topic is simply scheduling a lunch date next week. The robot goes running with my wife when I am away, with me along for the ride so I can still engage in dialogue and companionship with her even though half my attention is on a conference presentation.

In case this seems too far-fetched, note that in October 2011 Apple first released the iPhone 4S with Siri, a personal digital agent already on its way to scheduling dates and arranging dinners, all in plain, spoken English (Apple Computer 2011):

Ask Siri to do things just by talking the way you talk. Siri understands what you say, knows what you mean, and even talks back. Siri is so easy to use and does so much, you'll keep finding more and more ways to use it.

This digital agent is already more than a sophisticated voice-activated keyboard. Apple explains that it is smart enough to ask

you questions until it understands the instructions you are giving. Here is the birth of early agency:

And Siri is proactive, so it will question you until it finds what you're
looking for.

The Siri agent depends heavily upon web services, and combines
those web tools with GPS-based location information and all the
user's personal details seamlessly available through every smartphone application:

Using Location Services, it looks up where you live, where you work, and
where you are. Then it gives you information and the best options based
on your current location. From the details in your contacts, it knows
your friends, family, boss, and coworkers.

This last bit is sobering, pushing our conception of Siri from a
simple novelty to an agent with real power and armed with a
nearly threatening amount of personal information. Knowledge
of all my contacts, all my locations, and proactive agency: the
three combined make me wonder about privacy, data mining,
and of course just how embarrassing manifestations of any system bugs or viruses can be.

As artificial intelligence advances, such agents will become
only more sophisticated, able to better mimic my speech patterns and my interests, and even able to decide when and how I
should be brought into the conversation personally. The fan out
can increase with every advance in robot autonomy because my
robot proxies will need me less frequently. I become a puppet-
master who has an ever-increasing number of fielded puppets.

Taking this forecast to the limit casts a shadow on human-
human relationships. As my robots and agents conform to me
more closely and need my help less frequently, I become CEO
of a company where the company employees—my agents—
need me less and less. Soon I am strategist-in-chief, and nothing

more. Of course I can choose when I want to be present, and I can pick and fall into every experience that I deem most valuable to personally witness. In the limit, life becomes fragments of high-value experiences, with little time for the redundant, boring, or undesired.

I call this condition *attention dilution disorder* because it technologically instantiates nearly the same psychological handicap that is diagnosed today as attention deficit disorder. The one difference is that, in the eyes of many, the new ADD will be a desirable station in life rather than a condition to be treated.

To be sure, this new ADD will increase apparent business productivity, just as economies of scale yield quantitative benefits in everything from egg production to the garment industry. But consider the negative repercussions of such scaling: environmental damage, nutritional dearth, and increasing social inequity. ADD takes scaling to the personal, human level, and it is there that we could see our quality of life suffer under the strain of massive social fan out. By delegating part of our life story to agents and robots, we will sacrifice the depth of feeling born of an experience that firmly plants us in one place, with awareness and focus fully intact.

In *The Tipping Point*, Malcolm Gladwell argues that the behavior of people has great sensitivity to the context in which we live (Gladwell 2000). He also provides caution that the number of authentic relationships we can have is limited, and that communication technologies that provide massive reach outward do not really deliver a genuinely increasing social sphere. Our physical reach is set to explode outward, and I believe many of Gladwell's cautions will apply to this new form of fan out. We are in danger of slipping into an inauthentic, powerfully broad reach that gives us an even greater sense of personal power while stealing from us the very attention that keeps us centered and balanced.

From personal experience, I can say that this book exists because, to write each chapter, I took the time to take simple walks, chat with my wife, and just think. If I were living four lives simultaneously, context switching between them, and managing them as my own CEO, there would be no such book as this to publish.

5 Brainspotting

Guessing Game

Green Park, London, April 2231

"The wind helps."

"Wind? Look at that one. What do you mean wind?"

"It makes things more dynamic. Stuff happens around people. You look at how they respond. I'll show you, after a big gust."

"I think size is the giveaway. Like that guy. He's huge—really tall. What are the chances that, if he's a patch, the dude himself is really tall? In fact, yes! Check it out. He's walking with that woman, right? But look—every time they walk forward a few steps, he's in front of her, and looks over and hesitates till she catches up. Then, hah! Look at that. It repeats. Score. Every time—he's not used to long legs."

"Okay, that's pretty good, Dee. I'm impressed."

"I bet the guy is actually a foot shorter. He's never patched to a tall guy. How funny is that."

"You know what would be interesting? A gymnasium for patching practice. Wouldn't that make sense? You could have an obstacle course and other stuff—maybe forks and knives, whatever. And you get used to a patch before you go to a real location."

"I like it—but of course they'll charge for that, too. It's already too pricey—look at us—we do this for, what, twenty-five minutes?"

"Yeah but it's what the market can afford. Any cheaper and we'd be on waiting lists for a month."

"Oh. My. God. Did you see that?"

"Mr. Short-Tall Guy's woman friend?"

"Patch!"

"Absolutely definitely. That was awesome. She actually giggled. Out loud."

"Yep—looked right down her own dress and giggled."

"So that's our first man patched into a woman, for sure. Have we ever nailed that before, ever?"

"Oh crap. She's looking at us."

"Look at me. Bob, look at me, I'll see her anyway. Yeah, she's staring."

"I'm looking at you. Looking, looking."

Dee starts sobbing quietly.

"Wow. Dee. What's wrong?"

She winks briefly and keeps sobbing, then puts her head on his shoulder and whispers into his ear. "Problem solved. She saw me crying, figured we're just having a sensitive moment. Heh."

"Okay, you're good."

"Oooh. I have an idea. Check this out." Dee does the loudest imaginable sneeze, with a high-pitched trumpet that turns heads. Then she does two more, trying hard not to crack up. "That guy, Jeans. Bob, he never even glanced. I bet he's a 'bot."

"He is facing dead ahead. Walking pretty evenly. What if he's on a call?"

"Lips aren't moving at all. He's not nodding. I bet you he's a 'bot, fresh dead or just finished a patch."

"Wait—look, he's looking at—hey—he has a watch on his wrist. He's checking the time!"

"Damn. Busted. So he has to be natural."

"Who else checks the time. I think he's depressed. I bet he's walked here so much that nothing excites him, not even a crazy sneeze. He's a local."

"Check that out—he's just looking."

"At the tree. Weird."

"Okay—so more evidence. He's too patient. He's clearly not paying rent—no way a patch can just stand there like that."

"Look at us. We're just sitting on a bench."

"He's solitary. That's a giveaway, too. Natural, no doubt."

"And guess what, two-minute warning here."

"Shame. Hey—Dee—look at those two labs."

"No owner?"

"Yep."

"Their looks—since when do dogs look at each other, then tear off like that?"

"You think patches? To dogs?"

"I heard about it—a blast—you can swim, jump, all sorts of stuff."

"But you can't talk!"

"Actually—they worked that out. You can backchannel the talking—like an audio call—on the net. You mix it with the audio signal from the dog's body."

"Remix. Aha! Wow—now I'm convinced. Dogs just don't do that."

"And I know why. Because dogs can't read! I love it. Didn't anyone explain to them that if they read the signs in the park, it's a dead giveaway?"

"I don't think they care if someone notices. We're just the weird people watchers, is all."

"Okay—final warning coming in."

"I have an idea. The problem with London is it's Tourist-ville, so nearly everyone's a patch."

"Except for depressed wristwatch guy."

"Except for depressed wristwatch guy. But let's do somewhere where there are more naturals, so this is harder."

"Where?"

"Where. I know where. There is a sculpture park outside the Walker—Minneapolis."

"Deal. Can you make the reservation? Wait—I want to row! I heard there are great rowing setups there. Can you get us rower bodies and we'll row on the lake?"

"You know those things take balance, Dee? It's not like a paddleboat."

"Duh. And if we fall in, then we get to swim. Not so bad!"

"Geez. Okay. Two rower hunks, I'll look."

Giving Up

Boston, Massachusetts, April 2126 (105 years earlier)

The package itself felt old—like a prop from a time when materials were so cheap that objects were boxed, and boxes torn away. She took out the launcher and remote, picked a spot on her right forearm. The prick of the needle was painful and she stared at the spot. There was an expanding heat throughout her body—slowly. This takes time. After all, blood still takes two minutes to tour the body.

The old-time remote had a real button. Pushing it would have a satisfying feeling of selection. It actually moved—it had a spring and a moving part—and this was something she had not seen in years. The button required a real exertion—not a thought, not a wave of the hand, but a real connection to a made object. Then her thoughts turned to the result: death. What am I dying for?

She remembered first reading about robot control, and how upset she became. The nanos actually controlling muscles directly, not just manipulating chemicals in the brain. How far could this go? The first experiments on humans were clumsy, and this frustrated the budding scientist in her. No measurement of reaction time, consistency in motion, long-term effects. They simply demonstrated that the right arm can reach out and press a pleasure or pain button equally easily, no matter the sleeping subject's drug-induced torpor. There was so much excitement back then. News articles, new achievements every week. Riding a bicycle, typing on a computer, playing classical guitar, and then the big challenge: the ten-minute Turing Test. Watch Dave in a room for ten minutes. Is Dave in control, or are the nanos in control? But of course that was easy compared to real conversation—so much tongue control, vocal cord work that it would take two decades to master.

The nano control threw her for a loop; it really bothered her deeply, and everyone—her friends, her professors—was running headlong into nano studies. She had always had an aversion to the popular thing, and her own epiphany occurred during third-year university, when she read an old Asimov story and it hit her: she wanted to make mechanical robots rather than helping to perfect human robots. Years of mechanical engineering, electrical engineering study followed, at the top university still offering robot coursework.

Those six years were punctuated with one unforgettable argument that made her bitter. Her mother had a death policy but was an old-time recluse. What an odd combination—to set up a payout for the ultimate recycling operation on your body and yet, in life, to live alone and far from everyone. She was dead for days before her body was discovered, and it was much too late to harvest it for the human robot experiments. The insurance companies wouldn't pay—no upside in death, no death payout. She was the fool who argued, as a form of grief for her mother, breaking into regular yelling matches with the insurance people. There was no small print, just a vague clause that failure to harvest the body was material to the contract.

She did get her robotics doctorate and even landed a nice research position, but there was simply no funding for artifice—from government or industry. Humans were too easy to obtain, and the nano control research was progressing at light speed. All the research money, all the business money was thrown at nanos. Proposal after proposal came back to her rejected, always noting that if she had pitched the same research agenda but used human robots rather than mechanical robots she would have received funding in a second. And of course her whole point was to show that they didn't need to use humans, so she would have none of this.

Five years later, mechanically minded conferences had all but disappeared, so she was publishing in the robot conferences where everyone was working with humans. The provenance of robots was on her mind, and she spoke plainly of it. What place do the human robots have in our society? If you kill one, is it really only vandalism? What rights do they have? Must they be distinguishable from human animals? Only forty, she was already the curmudgeon, avoided by colleagues at conferences, never invited out for dinner where everyone would gossip about the keynote speaker's research results.

What pulled her out of her cocoon of research and frustration was the goodbye party for her old friend Sarah—from her very first year at university when they were just four very close friends experiencing all things new together. Sarah had done it—she had set up a goodbye party and had scheduled her death. Massive payments for twenty years—enough to buy anything, do anything—and now the time had come to respect the final entry in the contract and give her body to nano business. She sat down and talked with Sarah, and really listened. Sarah was truly happy. She described incredible travels, a full lifetime of experiences packed into two decades without worries, without the nonsense of having a boss, performance reviews—none of it. She asked Sarah, could she change her mind? Do people ever do that? Live high for twenty years, then change their identity and hide in Nepal? Tracking is too good, it turns out. And the contract is careful. The company actually owns you at the twenty-year mark, and receipt by death is totally legal at that point, so they can simply kill you to obtain you.

So she had to learn more and, after the party, she scheduled an informational interview with one of the biggest nano businesses, The Good Life—this one funded with a budget the size of California. The company made its pitch, much like what Sarah had described, then offered up a special that was just being announced: unlimited spending for ten years rather than the regular twenty-year deal. Unlimited. She pushed them on this and got to the details—it had to be personal spending (you can't give billions to your friends) and it had to be noninvestment (you can't buy gold and found a new bank), but it was for real. She asked them if she could fund her own research and they said, yes, she could do arbitrarily expensive scientific research for ten years.

That visit with The Good Life set her up on a mission. And that mission took her on yet another journey, through law school and specialization in jurisprudence all the way to her big case at the U.S. Supreme Court: *Rachel Ives v. The Good Life*. One dedicated soul in one corner, an entire army of infinitely funded lawyers in the other. The case was complex because technology had moved so quickly into business practice. One entire strand concerned motherhood. The legal ramifications of robot motherhood, from pregnancy to child rearing, had simply been unaddressed. Can the robot be a legal agent, even if purchased by a man expressly to carry and produce offspring, then act as supernanny? What happens to the robot mother-child relationship if the father dies? She set out a case arguing that the inherent inconsistencies in law and social dynamics suggest that we have to forbid robot motherhood until and unless we have full robot human rights equivalence. And of course business would never, ever allow for equal rights since that would destroy the economics behind the entire industry—then you're simply making more "real" people, which is uselessly uninteresting.

The second part of her case was about the entire death-selling business mechanism. It was the slippery slope she feared. Insurance companies were stepping in and offering year trades for long-term care of a loved one. It crept in at first in case of accident—we will pay all the medical bills for your spouse's terrible illness, and all you have to do is sign this slip that gives us your body if you experience accidental death. Later, of course, the terms became stricter, with a death date, and even bonus payouts for early death. Then the pawnbrokers got involved, offering up quick cash in trade for future exclusive rights on handling your death terms. Then prison wended its way in, with families asking lifetime sentence convicts to sign a contract. One final deed—a good deed—would make the convict able to provide for the entire natural life of his children. Her argument in court was that the commodification of death had changed the social calculus so severely that fundamental human rights were in question, and therefore the death upside needed to be invalidated. Of course The Good Life did have one huge ace in its deck— government studies showed a total collapse of the economic system if the robot business were interrupted—it was too big to shut down.

The Supreme Court's final ruling was ambiguous and disastrous. Sections read like gibberish:

The individual and societal pursuit of liberty per se accords decision-making authority to each person, subject to the rules and regulations governing civilized life. But liberty applies only to living souls, and the use of former bodies as vessels for nanorobotic systems, whether remote-controlled or autonomous, in no way impedes liberty or freedom for those who live. This is no more than a simple extension of any use of dead matter, be it oak wood used in furniture or a lock of hair used in a modern art piece or in a wig. Furthermore the form of manufactured artifacts cannot be regulated on the basis of how similar they look to natural products. However, formerly alive humans are not physically manufactured, but we view the control of these bodies by nanorobotic distributed control systems occupying the brainstem as a form of manufacturing, and therefore we define the entire system as manufactured. While the sale of organs continues to be illegal, note that the defendant's business methodology involves the sale of material and not living tissue. It is true that with nanorobotic control the body can continue to function biologically even after eradication of consciousness but human life as defined by cerebral brain activity is absent and therefore sale or purchase of the body falls in the same category as sale or purchase of a fresh-frozen fish.

Now she was home, with a most unconventional exit before her: a death kit without a death policy. She had no children and no friends. No one to benefit from her death. Why die? Because she had a broken heart. She believed in humanity, explored humanity, fought for humanity and utterly lost. No glimmer of hope. So, she pressed the button, and as she hoped it had a satisfying, solid motion down, accompanied by a beep that she did not expect.

The nanorobots stand at attention at every synapse in the brainstem. They found their way there by injection and communicate with one another in perfect synchrony. With the press of the button, the message spreads throughout the network: time to take over. First, a synchronous release of nerve suppressor. No more consciousness. One quarter second after this death, the nanorobots take over, restarting the heart, the lungs, the endocrine system, the digestive system. There are two things that you can see if you are looking at Rachel's face really closely when all this happens. First there is a change in expression—the muscles go from an anxious look to a totally flat appearance—neutral, somewhat like in sleep. Second, there is something that changes in the eyes. A sparkle that goes away.

Next stop: the death certification center. The robot stands up and begins the journey with a walk to the train. There will be a brain scan and EEG to check for lower brain functions. There will be finger, voice, and eye matches to verify identity, and then the death will be registered. Rachel Ives will be officially dead and a new serial number will be assigned to the robot that is her body.

At the train station, there are a couple of familiar glances as the robot sits in an empty chair. Rachel Ives was famous, after all—the only person to have ever fought The Good Life publicly. A young girl crosses over and sits next to the robot. "I am such a big fan of your writings and your speeches, Miss Ives. You have really opened my eyes and I just don't know how to thank you. It is such an honor to see you and I mean, thank you for being who you are. It means so very much."

Robots are not humans, but the awkwardness of such encounters is well documented in their shared experience. To inform someone on the spot of death— that the person they think they're talking to is gone—is too awkward. There is an elegant alternative that ends up playing out in many such chance encounters, on the way to the death certification center. A smile, a thank-you, a closing of the eyes sleepily to punctuate the end of the conversation.

• • • •

The previous chapters have proposed robot futures in which incremental technological advances in robot perception, cognition, action, and human–robot interfacing change the way we relate to our social world. But there is a more disruptive robot future that I explore here. What if robotics succeeds best by exploiting human biology—using our joints, muscles, and energy systems rather than by inventing new mechanical forms and materials? The biological efficiency of animal systems stems from millions of years of evolutionary optimization. We have suppleness, dexterity, and an efficiency of motion that are unparalleled in the machine-built analogs engineers have created so far, and it is not yet clear whether our fuel cells and electrochemical robotic inventions will ever be able to approach biological systems in terms of robustness, dynamics, and energy efficiency.

If robots are to fully exploit biology, what will that look like? To answer this question, first we must look at the state-of-the-art coupling of robots with the natural body. The simple interaction of robots with biological systems is not new. Already, one of the major success stories of robotics is proving to be its application to human quality of life in the form of robotic prostheses. Active ankle joints in prosthetic legs have enabled amputees to walk up and down stairs far more easily than before (Au, Berniker, and Herr 2008). These robotic lower legs have power supplies, electrically controlled joints, and accelerometers that sense different types of walking gaits and actively manage joint resistance or flexion to work in concert with the biological portions of the leg, such as the knee.

Other researchers take a different tack, developing robot *exoskeletons* that wrap around human legs and provide extra strength for walking and using stairs. The original Department of Defense funding for this research stems from the need to have soldiers carry ever-heavier packs during war: is there any way a robot can enable a soldier to carry a pack that is fifty pounds heavier without feeling the extra weight (Kazerooni 2012)?

But early successes in exoskeleton research have multiplied, which is leading to a very exciting revolution in mobility for the elderly and the physically challenged in nonmilitary applications. In upcoming decades, individuals who today use wheelchairs will begin to adopt walking, robotic exoskeletons instead. These robots will start out as controllable devices at coarse levels of fidelity, like an electric wheelchair: they will command the legs to walk, turn, run, and sit using a simple interface. This will already be a massive improvement over wheelchairs because users will have far more comprehensive access to our built world. Moreover, the psychosocial empowerment of standing tall and making level eye contact with peers means that the world will begin to treat these individuals with more respect and equity as is their due. This may turn out to be one of the most fulfilling examples in history of technology transfer from Department of Defense funding to civilian society.

There will follow a natural desire to couple the motion of such robotic prostheses ever more intimately to the muscular, fine-grained desires of the operator. Experiments today already demonstrate detection and processing of brainwave signals from the scalp to control the motions of a robot arm (Bell et al. 2008). As the brain-machine interfaces improve, higher-fidelity control of robotic prostheses will advance rapidly.

How far can innovation in brain-machine merging take us? Two areas of long-term technological progress might make the ultimate human–robot couplings a practical reality: interface production and interface control. Interface production is the physical problem of coupling digital signals to biological, living nerves. Interface control is the computational problem of making sense of neural signals and mimicking those signals to actively control muscles.

The first step is to create physical interfaces that will measure signals in all the relevant synaptic neural locations. Without this direct measurement, the robot controller, however intelligent, is guessing indirectly at human intention. These synapses, the connections between neurons, are where messages are passed that activate and control the muscular systems responsible for movement.

The physical interface is where nanorobotics can play a role, by forming massively large robot colonies, within the body, that yield an interconnected human–machine interface. To measure signals effectively, nanorobots will need to enter the body, migrate to synaptic interconnections between neurons, and then measure values in real time. To move into these positions, nanorobots will need to be incredibly small—possibly molecular. Red blood cells are less than 10 microns in diameter, and to be able to position themselves without harming the body, each nanorobot will likely be similar in size. Even at this small size, red blood cells deform to enter the smallest capillaries; similarly, the size and flexibility of nanorobots will be a massive engineering undertaking.

I would hypothesize three basic delivery mechanisms for transporting nanorobots into the human body: direct injection into the bloodstream, inhalation for pulmonary transfer, and transdermal application (a skin patch). The materials would need to be sufficiently inert to avoid immune response; they must be carefully designed to avoid congestion and clotting in smaller arteries. They must also be able to travel through arterial walls without causing cell damage.

The total number of nanorobots that would be required for a human interface is staggering. Each human neuron can have 10,000 interconnections, and it is these interconnected signals that robots need to access directly. Even in the limited cross-section of the spinal cord or brain stem, signals across one hundred million neurons would translate into more than ten trillion synaptic measurements—which suggests many trillions of nanorobots would have to be deployed, just to comprehensively patch into a single human interface.

Once a nanorobot network can measure all relevant interconnections, the next challenge is to interpret those signals correctly and then use that interpretation to effect real-time control. This is a machine learning challenge that computational researchers will enthusiastically surmount. The artificial intelligence system would need to learn patterns of signals and understand, over time, how these patterns correspond to the right muscular control outputs when combined with environmental sensing provided on the robot side. As difficult as the signal processing and interpretation problem may seem, especially when facing trillions of new values many times a second, these are straightforward compared to the raw engineering challenge of manufacturing the physical nanorobot network.

Prior chapters predict robot advances by extrapolating incrementally from existing innovation trends. That is how we can imagine a few decades forward with some confidence. But this chapter considers a robot future that humanity cannot reach incrementally. Scientists would need to overturn major, completely unsolved problems to reach this destination. For one, physical fabrication of trillions of micron-sized robots is impossible today. Energy storage and harvesting for such robots is also entirely unknown. Even if each robot could function individually, networking trillions of robots together (whether chemically or through electromagnetic signals) would require whole new innovations in communications technology. One hundred years is my best estimate for such revolutionary advances, but an honest appraisal in this case balances on the question of if, rather than when, such advances will occur.

If and when a nanorobot mind–computer interface does become reality, it will do much more than simply measure neural synaptic values. Humans are learning feedback systems—we are constantly adjusting our muscular control based on real-world experience, never acting without perceiving. The same nanorobotic technology that measures trillions of signals will start providing feedback to the synapses as well—supplying impulses that represent the signals that the prostheses ought to provide if they were biological. This is an even harder AI challenge, in that feedback strategies must learn to match what neural networks expect. But when this works, a prosthetic hand will begin to provide the sort of feedback to the brain that successfully fires the imagination of an artist who hopes to sculpt in clay again, or the musician who needs to feel the membrane on his drum to play with emotion.

Once you take both interface production and interface control to the limit, a new space of possibilities opens. Suppose nanorobots can present and measure across the entire breadth of spinal cord and optic nerve interconnections. Suppose also that AI analysis and control progresses to the point that this entire collection of signals can be interpreted correctly and further reformulated to provide feedback to another brainstem similarly connected to another nanorobot cluster. At this point, the nanorobot colonies can become a new lingua franca capable of translating between one brain, over here, and another body, over there. Dick can patch into Jane's brain stem, feeling the perceptual inputs of Jane and controlling Jane's muscles directly through the same interface. Where will Jane's mind be during this form of remote control? Strangely cut off from her own body's sensors and motors. Perhaps she will be patched into another person, or robot, or simulation.

During a patch there is also no particular reason why the brain and the body need to belong to the same species of animal. Computers can perform real-time *transformations*, translating musculoskeletal action commands on your part into the musculoskeletal analogs in a common eastern gray squirrel. With practice, you can run through a tree canopy, jumping from tree to tree, feeling the rush of moving across land without ever touching the ground. You can patch into a python snake, with a computer translating your desire to make forward progress through a rubble pile into the complex physical gaits enabling the snake to easily tackle unstructured environments. You can look for victims of an earthquake in the smallest crevices, *as a snake*. Of course there is nothing stopping engineers from also putting robotic devices on such a dead-headed snake, for instance giving you the ability to deploy a small medical robot-pack to measure pulse, blood pressure, and provide hydration to a victim.

The plug and play of brain and body surfaces a welter of ethical and practical consequences involving two basic concepts: *identity* and *accountability*. Technological progress has already been challenging classical notions of accountability for decades. Technology tends to increase the complexity of systems along several axes—the number of people partly responsible for a newer product is ever larger; the amount of software in new products dwarfs earlier systems; the interface used by the operator to control the product becomes more intricate. All these axes of complexity make the resulting system errors less clearly understandable and less accountable, with no one ever directly or solely responsible for the behavior of a complex robotic system.

Technology ethics and design courses frequently study the tragedy of the Therac-25 to understand how much can go wrong when poor design, incorrect training, and simple errors are compounded (Leveson and Turner 1993). The Therac-25 was a radiation therapy machine that provided focused radiation to cancer victims to destroy malignant tumors by rapidly moving a high-energy radiation beam. The nurse-operator of the machine would configure the machine for a customized treatment pattern, then launch its autonomous radiation therapy mode. In the rare event that the operator entered the mode incorrectly, then backed up in the interface and corrected the entry within eight seconds, the machine would configure to an incorrect internal setting, where it would deliver one hundred times the intended dose of radiation, inducing massive pain in the patient and, eventually, killing patients through radiation sickness.

Many aspects of the Therac-25 therapy process are partially to blame for this. The interface was poorly designed, making incorrect data entry easy. Training for the operators was lightweight, and the nurses afforded the expensive, fancy machines more authority than the machines deserved. When the patients complained of pain during the procedure, the nurses would discount this complaint because the machine indicated that all was well. The fatal software flaw—that a specific series of keypunches could put the machine into an unknown mode—was not discovered early because systems testing was not comprehensive enough. Bad design, poor testing, inadequate training, and an inappropriate operator-machine relationship—all these factors combined to cause lethal errors. In such a sophisticated scenario, there is plenty of room for blame in all corners, but there is moreover no single point source of blame—no accountability to any individual who deserves responsibility.

As technology advances, new innovations induce even newer relationships between machine and operator. Telepresence, for instance, further complicates the relationship of individual human decisions to outcomes in the natural world. The Predator drone is a militarized unmanned aerial vehicle controlled by an entire team of operators thousands of miles away using highly sophisticated software. Furthermore, the operators are themselves located within the hierarchy of a military command structure, with limited access to information regarding their mission and its contextual background. When a remotely operated Predator drone reacted to a wedding party on an Afghan mountaintop, killing members participating in a marriage ceremony after a celebrant fired a rifle in the air, where did responsibility for the consequences lie? The operators were not physically present and instead depended on a limited perception of the theater based on software analysis. Furthermore, the system combined human control with levels of adjustable autonomy; at any given time, the drone may be under the guidance of a human, a team of humans, or a piece of control software that is making second-by-second decisions about headings, speeds, and potential threats. In such a scenario, accountability for the consequences extends among an operational team far removed from the consequences of their decisions, and very limited in their understanding of the perceptual limitations of a complex robot, and a command hierarchy similarly undereducated about the perceptual and control failure modes of the technology they have embraced (Singer 2009).

A recent news article in the *Washington Post* reports on drone maneuvers at Fort Benning, Georgia, in 2010, where roboticists have programmed an unmanned aerial vehicle to fly autonomously, use on-board cameras, and computer vision algorithms to search for an object on the ground, match it to a desired target, then automatically fire on it (Finn 2011). The article goes on to explain that the eventual goal is for drones to have a database of enemy images and fly above a battlefield, looking for the desired targets with face recognition software. Following an image match, the drones would then make the autonomous decision to kill on their own. This is a poignant example because, from some distance, it may appear reasonable to a nonroboticist that drones can do face matching and then make lethal decisions. But to any roboticist aware of state-of-the-art computer vision and face recognition, the concept is absurdly out of touch with the reality in which face recognition can easily find false matches using anything from the rear end of a cow to a poster with the target's picture pasted on it. In this possible future, you might be able to kill your enemy by simply taping a picture of a known terrorist to his front door.

As we extend the concept of robotics and adjustable autonomy to remote brain-body connections, the predominant impact humans have on the world is mediated by imperfect, rapidly changing technology. The primitive causality of action to effect becomes a rarity when every perception has been interpreted by machinery, every action has been translated and scaled by software, and every human relationship has been forged on the anvil of machine interaction. Recalling that all software has bugs, that all produced systems contain errors (some of which are never discovered), then I wonder: how can a system so replete with layers of imperfect technology ever afford the casuistry of accountability?

If accountability is endangered in this possible future, then identity faces an even more certain threat. People can already step outside their usual locale and even outside the mores of regular life: video games give players the chance to engage in combat or sports that are unthinkable in real-world bodies. But the whole self is always playing a role—the body is staring at a video game screen, and physical hands are giving rapid commands to a game interface. Even telepresence systems thirty years hence—described in chapter 4—do not genuinely disconnect the mind from the body. The sensing and control of that remote location always flows through the body—hand muscles tap buttons and keys as well as joysticks; eyes and ears provide sensing feedback thanks to glasses or screens and earphones.

The cues of the here and now are ever-present so long as the body mediates the connection to another place or another universe, and this suggests that the self can retain a basic level of grounding. Someone can always put their hand on your shoulder and get your attention *here*—even if you're playing a video game in Narnia, conducting a tele-meeting in Nepal or rock climbing in the asteroid belt. Each adventure can feel real, but the immutable fact is that your mind and body are together controlling that experience. Each telepresent episode is like a fancy dress party—the experience can enable you to explore a different identity, but you also will also have your own body as a foundational cue. Remove the costume and it is still you underneath, nothing more.

But what happens when we disrupt the nature of this reality? What if, TRON-like, you wholly jump into another world, whether simulated or just geographically disparate? Your own body, your own physical being, ceases to mediate the experience. You become a thinking system embedded in a new body and context. Your hand is a new hand; your face is a new face, through and through. As patching becomes ever more sophisticated, a patch can feel only more natural, and this means that your remote experiences, entirely unrelated to your actual body's status or physical location, can become just as formative as your proximal experiences.

Your current notion of identity can lose meaning if your true physical context changes at any time. You become defined, not by an identity stemming from your physical and psychic self, but by your history of actions as whomever and wherever. You are an experiential participant—and that participation is what defines you. In this extreme analysis, attentiveness to the physical body can diminish, since no particular corporeal form is especially unique. Of course, since your own body would not feel any more natural than any other body, your sense of connectedness to that original vessel will also attenuate.

The great irony of this brain-swapping possible future is that artificial intelligence research has labored for decades to create human-level AI: thinking machines that can be the perfect agents without bodies. What if the future of human experience begins to reduce us natural human beings to bodyless agents? We might just invent the perfect agent after all, but not by creating it from bits and bytes in a computer. We may accomplish the AI dream by stripping humans of so much singular identity that people are reduced to mere agents.

Of course, in the brainspotting truth-is-stranger-than-fiction department, a whole litany of clichéd Hollywood scripts become achievable as well. Two persons swap bodies and experience one another's daily lives. One person formulates the ultimate synchronized swimming team by directly controlling two bodies simultaneously, modulated only by a computer interface that performs all the right transformations for variations in individual body mechanics and musculature. Two brains patch into the same body, sharing control or splitting up the body—*You take the hands. I'll take the legs.* Desperate individuals rent out their bodies for money, confining themselves to their brains during a one-hour rental, and witnessing (if they are still receiving sensory inputs) their body acting under another's control. *The Matrix* meets *Being John Malkovich*.

For even more surrealism, go further back in time before Hollywood, to the Brothers Grimm. What about that prince who turned into a frog? The interesting question becomes, who gets to decide when to turn off a patch? If the wicked witch is in charge, she can patch you into a frog, once you've inhaled the nanorobots. And now you're a frog, until she decides to cancel the patch and allow you to feel and control your own body again.

Much of this begins to feel remote enough that worrying about such a future is frivolous. But this possible future is much easier from an engineering perspective than the mind duplication extolled by singularity adherents, who often estimate their ideal future to be only a few decades away (Kurzweil 2006). Their argument often goes like this: if scientists can create a full-scale model of the brain, by mapping every neuron and every synaptic connection, and then measure the actual neural transfer functions inside a specific person, then science can copy the exact brain *state*, and thereby the unique "consciousness" of that person, into a model that is implemented in a computer rather than in a carbon-based, biological life form (Moravec 1990). This duplicated, simulated brain would be, by this way of thinking, as self-conscious as the original from which it was copied.

But mind simulation and duplication is even harder than the nanorobot interface I have described: it requires scientists to be able to model the entire human brain computationally. This is a tall order, by nature of the complexity inherent in the brain. In contrast, the nanorobot interface needs no knowledge of how the brain works—just the ability to interrupt signals from the brain en route to muscles, and the ability to report back signals from nerves en route to the brain. Even if successful, mind duplication ceases to really be duplication the moment a simulated being comes to life—the new being is instantly unique because every experience it has henceforth pushes it further away from that common ancestor, even though its background and memories were identical to that of the original person up until the point of duplication.

I would not personally hazard a prediction for when scientists might achieve full-scale brain modeling and mind duplication—the number of years I would estimate would be so large as to be meaningless. But if mind duplication were possible, suffice it to say that even greater threats to the concepts of identity and accountability, not to mention human rights, property rights, and suffrage, would undermine any system of societal law in place today. In short there are many, many good reasons to avoid copying our minds.

The real problem with this speculative line of discussion is that talk of mind duplication and immortality is not just too distant and too frightening—it is also desperately exciting and inspiring. Stories of technological utopia decades or centuries in the future do not solve the problems we face today, nor do they help us guide the development of technology in the near term for the best possible human good. The near future of telepresence, robotics, and communications technologies threatens to distract us, dehumanize our interactions, and erode our personal freedom and choice. The true challenge we face is in charting a new course that instead celebrates and nourishes individual well-being, accountability, and societal equity. To chart such a course we should become more deliberate and considered as we imagine and design technologies that carry us forward.

6   Which Robot Future? A Way Forward

Robotic technologies will confer new powers to us. We will observe on a massive scale and automatically respond; we will interact remotely, explore dangerous or distant spaces, invent new toy species, and even be telepresent in many places at once. Perhaps one day we will be able to assume novel physical shapes and construct experimental identities through which we explore and engage with the world. These are all forms of technological empowerment, and prior chapters have discussed *who* might be the likely beneficiaries. We started by speculating how corporations will continue with ever-greater acuity to optimize information gathering and interactive marketing with unprecedented personalization and targeting of tailored messages to manufacture desire. Robotics also empowers the military, enabling new war fighting at a distance, and new regimes of overt and covert military action that blur the distinctions between peace and war. So, from the perspective of institutional bodies, robotics provides great increases in all forms of power and influence: the institution can perceive more comprehensively everywhere; it can act with broader impact and, at the same time, with more customized local impact anywhere and however it chooses.

But the newest robot technologies also empower individuals without the usual normalizing constraints of social accountability. As robotics becomes simultaneously more accessible and more capable, anyone's ideas, from creative to extreme, can move readily from concept to actual walking, talking robot reality. Robotics will empower solitary individuals and self-organizing groups to graduate from expressing radical ideas using the Internet to forging physical expressions of those ideas through embodied mechanical action. Anyone with a modicum of technological know-how and access to online open-source communities will be able to build a robot that has the potential to push buttons in the physical world, and this will feed antisocial applications, even vigilante robots like BumBot (chapter 2), as readily as altruistic inventions. The Internet is already a place where ideas and opinions, however marginal, can attract enthused supporters and self-reinforce almost any belief system. Robotics will take this same dynamic into the physical world we inhabit.

The prior chapters strove to provide examples of how robotics may impact human interaction in questionable and complex ways. The recurring theme of each chapter is the specter of empowerment gone wrong. Institutions benefit, but the problem is that their goals never align perfectly with those of society as a whole. In fact further empowerment of corporations can cause *disempowerment* in communities as new technologies asymmetrically and opaquely confer the power to shape information and manufacture desire.

What if, instead, future robotic investments and innovations were designed more explicitly for our communities rather than for institutional or corporate interests? The idea of new information and communications technology (ICT) directly benefiting the people is no pipe dream. We have seen mobile phone and social networking technology change the world by surfacing new forms of documentary capture and communication (Deibert et al. 2011). Twitter, microblogging, and online video sharing have supported protests and revolutionary actions, and exposed police action even in the most tightly controlled countries. In one year we have witnessed successful democratic upheavals in authoritarian states where change was previously unimaginable. This is civic empowerment that makes the heart sing—citizenry able to capture, to share, and to act on the most essential human rights.

New robotic innovation could also play an important role in social change because it empowers communities with data and action, enabling positive social change just as mobile phones have done. The price point of robotics is falling dramatically even as the lowest-cost robotic sensors improve in their accuracy, battery life, and volume availability. Robots can measure and push back on the world, and because their components are becoming so inexpensive and easy to replicate, a truly useful robotic technology can scale more affordably and more quickly than ever before from a research lab to a local pilot to international distribution.

But for salient broader impacts and human concerns to emerge, we must find ways to put communities, whether defined by geography, affinity, or practice, at the very center of a new technology design process. One major roadblock to a design-for-good approach is how the research agenda is capitalized in robotics. Funding flows from industrial and military sources that have specific, self-serving criteria for innovation and impact. The agenda is set by the availability of money, and so the holders of the purse have disproportionate power over the direction of our robot future. A possible solution lies in creating greater funding balance by ensuring that socially aware funding sources encourage real community-centered design in robotics.

The National Science Foundation and other national funders already request that applicants describe the broader impact that the proposed work may have on society. But, too often, these impact statements play no real role in funding decisions. I have served on funding boards and I have seen technical content dwarf social impact in the eyes of nearly all the reviewers in the room. This needs to change; broader impact must become an authentic criterion for funding, at least on an equal footing with scientific merit and technical innovation.

Private foundations and philanthropy have been leaders in providing funding that aligns with community needs, tackling research focuses such as gender equity for technology education, programs centered around social justice efforts for combating poor health outcomes, local pollution, and energy inefficiency in neighborhoods. These foundations must take community-centered robotic innovation even more seriously, insisting that their grantees engage locally, model the needs of true end users and the ecologies they inhabit, and then innovate explicitly for positive impact. This will demand long-term, cross-disciplinary partnerships among communities, social scientists, designers, and roboticists. This vision will also require socially minded funding sources to direct so much substantial funding at robotics that the realization of community-positive robotics can be a genuine alternative to existing robotics funding practices.

At the research lab level, we need a healthy ecology of diverse approaches to meaningful technology design. Funders' goals vary richly, and if an equally diverse group of research labs can all strive for mindful, broad impact, each in their own unique way, then we have a chance at changing how the robotics research life cycle moves from ideation and funding to deployment.

At Carnegie Mellon University, I direct the Community Robotics, Education and Technology Empowerment (CREATE) Lab. It is one experiment in socially focused robotics development, and it has succeeded because we have been able to cultivate local foundation and NSF funding to support our community-centered robotics innovation focus for more than a decade. Such a long-term community commitment can only flourish with strong organizational support—in our case both the Robotics Institute and the university's administration have played critical roles in supporting CREATE's unusual fundraising agenda.

Community-centered technology vision demands a multidisciplinary approach, and so our lab brings together experts in learning sciences, psychology, design, engineering, and robotics with international partners including National Geographic, NASA/Ames, Google, UNESCO, and Science House Foundation. We collaboratively unearth local societal needs by working directly with community organizations as partners, not as research subjects or target audiences. We introduce robotic technologies to them in workshops, then employ participatory design techniques to elicit ideas and speculation about how they would imagine using robotics to respond to their most pressing local needs.

In two of our community studies, the number one concern that citizens aired revolved around transportation and several facets of its locally felt impacts—congestion, pedestrian safety, air quality, and noise-related pollution (DiSalvo et al. 2008; DiSalvo et al. 2009). The need, which we call *traffic calming*, is one that is felt in neighborhoods around the world. A friendly and quiet neighborhood on the outskirts of an older central business district grows with industrial development and sprawl into a major thoroughfare for commuters and trucks, and this completely changes the character and the conditions of the neighborhood. Traffic snarls make local trips painful; traffic density leads to noise and air pollution that in turn is related to greater asthma cases in schools and households unlucky enough to live near the high-traffic streets. The commuters, in turn, know little of the local neighborhood, treating it as nothing more than an annoyingly slow route to their destination.

Neighborhood groups want to calm traffic by creating empathy between commuters and communities, emotionally connecting the commuters to their surrounds. They have proposed and prototyped numerous robotic devices in our workshops to this end: interactive signs that remind loud commuters to quiet down because real people live here; smart boards that propose alternative routes by demonstrating commute times and real-time traffic conditions graphically; kinetic sculptures and interactive messaging devices that sense and respond to levels of air pollution, taking the invisible and making it tangible so that drivers see how they are contributing to local conditions. One neighborhood even proposed and designed a *radio robot*: an automated local radio broadcasting system that would detect traffic density on their narrow bridge, then broadcast targeted stories about the local community, its history, news, and businesses on Main Street—all to get the drivers stuck in bridge traffic to realize they are crossing a valley and passing through a community that contains a real, living neighborhood worth visiting.

Environmental degradation is another major issue that frequently surfaces in discussions with local communities. Water tables are dropping and becoming salinated due to overuse and industrial activity. Particulate matter in the air near power plants, road systems, and light industry goes undocumented while local communities attest to its ill effects. Rivers becomes tainted by acid from abandoned mines, polluted with runoff from high-volume agricultural practices, and contaminated by countless water treatment plant failures (Lerner 2010). Local communities witness poor health consequences, but rarely prove a watertight causal relationship and consistently fail to win relief. New community-centered techniques integrate the local experience of residents with professional techniques normally reserved for scientists who impact regional and national policy. In *Street Science*, Jason Corburn describes success stories from the Greenpoint/Williamsburg neighborhoods of Brooklyn, where residents have joined forces with professionals to make meaningful local change (Corburn 2005).

Robotics has an important role to play in the future of environmental street science because it can create tools that enable communities to collect data comprehensively, visualize it convincingly, and advocate more effectively. Low-cost air quality sensors, water quality data loggers, and health monitoring tools are already in our technology development pipeline. As citizens' ability to measure, map, store, and display environmental degradation becomes commoditized cheaply, communities can adopt and observe their own land, air, and water in a new technologically data-rich way. They can monitor their ecosystems comprehensively, with more frequency and spatial resolution, screen for outliers, demonstrate statistically significant evidence of causality, and make strong cases for business and regulatory change based on compelling evidence. Communities would be able to make the credible data-driven, evidence-based arguments of science that, to date, were reserved for scientists, licensed technicians brought into policy discussions by specialized technocrats and corporate experts. Robotics for community empowerment offers an exciting prospect for renewed democratic action because it enables local actors to reclaim power in a world where levers have been sucking power away from local citizenry: extremely concentrated wealth, corporate interests, technocratic specialization, short-sighted government politics, and the digital divide.

New technology is easily misunderstood, and first uses of innovation can greatly guide its downstream applications by narrowing peoples' imaginations and sense of personal invest-ment (Gieryn 1999). How can society better comprehend the consequences of the robotics future and the ramifications of its uses? And how can society meaningfully engage in designing the robot future in a proactive, inclusive way rather than simply witnessing its arrival as an observer and marketing target?

To create an engaged citizenry, we must also greatly improve basic technological literacy. The public needs to understand the outlines, dimensions, and affordances of oncoming robotic innovations well enough to participate in an informed discourse regarding their possibilities and consequences. Institutions that are involved in civic education can make great strides in this regard, from media, academic press releases, and forums for pub-lic engagement, to local science centers, adult-education pro-grams, and formal schooling. Today new robotic innovations are being announced using frames that emphasize technology as a savior, focusing on the wonder of invention but paying scant attention to its full consequences—boundary cases of applica-bility, failure, and ethical ambiguity. If robotics researchers and inventors bring a critical eye to broader impact analyses of their claims, if the media communicates not only the potential but also the limitations of new robotic innovations, if assessment efforts consider the ethical ramifications and unintended consequences more consistently, then informed communities can build the yardsticks with which new innovation can be more meaningfully subject to consideration, evaluation, and debate.

Another essential factor for responsibility lies with government: playing post hoc legal catch-up to the results of emerging technology fails to provide thought leadership on issues of accountability, identity, life-cycle analysis, human rights, and well-being. Every new telepresence and autonomous robot system will challenge the interpretation of our existing body of laws. Cars that drive themselves will crash in unexpected ways; robots caring for a child or the elderly at home will sometimes fail to notice the obvious; telepresent systems will be abused and cause mental distress to distant victims; all manners of robots will be used for crime and malicious acts in as yet undiscovered ways.

Instead of reacting case by case to new loopholes in law discovered by ever more ingenious machines, our system of jurisprudence must proactively gather the expertise and wherewithal to predict our robot future, debate the most critical issues of safety, accountability, equity, and quality of life, and create a viable legal framework for this century. Not only would such an exercise provide guide rails for future robot engineers and businesses, it would also catalyze a public awareness that we are entering an uncharted space but are girding ourselves with the knowledge and moral authority to make sense of our future.

Finally, a tremendous responsibility lies back with the academic world driven by funders and researchers—of course, this is the community that I know best. Robotics research has mainly been motivated by the desire to advance the state of the art in robotics. Let our robots see more completely, walk more efficiently, and think more quickly. The cultural impact of research—how each new technology will positively affect people—is often a distant concern at best. In such a world, innovation has value inherent unto itself. This works well in fields that unlock basic knowledge over timescales of decades and centuries, like cosmology and evolutionary biology. But robotics and artificial intelligence are now much more like molecular biology, medicine, and nanotechnology: the arc of impact from robotics has turned sharply inward, and the effects of researchers' inventions are now likely to be realized in their own lifetimes.

There is an increasing gap between how we train and mentor robotics innovators in academia and what basic skills they need in order to improve our world by plying their trade. We teach innovation, and they also need methods for problem finding with community engagement. We teach optimization, and they also need a history of morality, ethics, and law. We teach engineering and they also need guidance for public communication. A sea change in how a roboticist comes of age is warranted.

The next generation of engineers will make stories of magical realism come true—they will need the social, ethical, and moral tools to have the best chance at improving our quality of life rather than degrading it. I would advocate for a re-visioning of robotics and engineering curricula. This requires universities and professional engineering organizations to come together with the courage and willingness to chart the basic skills that every robotics engineer must have. Codes of engineering ethics and medical ethics already exist in other fields—robotics' impact will only grow, and so it is time for this field to mature from an unconstrained playground to a more formalized learning ecology like medicine and civil engineering. With the right effort we can more closely align the art and science of robot engineering with methods to guide the evaluation of human impact that will be so important in the ever-faster dynamics of change in the future.

Robotics is becoming a potent force, but, like much of technology, it has no innate moral compass. It is destined to influence society, and I believe the early adopters are already apparent: corporations, militaries, governments, and a privileged band of technically savvy individuals. What is missing from this list is the interests of citizens and local communities, motivated neither by power nor by economic value, hoping to contribute to a sustainable quality of life. Our challenge and opportunity lies in becoming the vanguards of ever-better robot futures, and this means we must bend the lines of influence that robotics will forge. If we succeed, we make an alternative vision into soaring reality: robotics as new, interactive media for making local change; communities empowered to measure, problem-solve, demonstrate, and act to improve their conditions. In this possible robot future, the robotics revolution can affirm the most nonrobotic quality of our world: our humanity.

now I see where
this revolution is leading
To the withering of the individual man
and a slow merging into uniformity
to the death of choice
to self denial
to deadly weakness
in a state
which has no contact with individuals
but which is impregnable
So I turn away
I am one of those who has to be defeated
and from this defeat I want to seize
all I can get with my own strength
I step out of my place
and watch what happens
without joining in
observing
noting down my observations
and all around me
stillness
And when I vanish
I want all trace of my existence
to be wiped out

—Peter Weiss, *The Persecution and Assassination of Jean-Paul Marat as Performed by the Inmates of the Asylum of Charenton under the Direction of the Marquis de Sade*, 49–50

Glossary

3D printing A rapid prototyping process by which material is added in layers, often using plastics that are heated and melted on, to create a three-dimensional model. In robotics, 3D printing can even be used to create structural elements that are bolted or glued together to create robot parts such as a chassis or manipulator fingers.

Adjustable autonomy An architectural consideration for robot control that embodies the notion that robots should be able to act on their own (autonomously) whenever possible, but that humans should be able to take gradual control of the robot along a sliding scale, from providing strategic oversight to manually and directly controlling the robot's joints.

Agency In design, philosophy, and human–robot interaction, this term has special meaning in denoting that a made object is demonstrating key aspects of decision making and enactment of action that we normally associate with humans. Another way of thinking about agency is in terms of intentionality—the idea being that robots with agency show an authentic intention in their behavior.

Analytics A methodology for gathering behavioral statistics for an Internet site to create an understanding of how people use the website, and a report on users' demographics.

Artificial intelligence (AI) In this book we use AI to denote the study of cognition inside software—that is, how scientists pursue the creation of decision-making software systems that can perform human-level operations such as social interaction with people.

Cognition   Cognition represents the AI-level decision-making and control processes on the robot that must choose how to interpret the sensor inputs of the robot, how to make sense of these in context, and finally how to decide what to do in order to push back on the world and enact change.

Computer vision   This subfield of computer science and robotics studies how the simple images returned from digital imaging systems like digital cameras can be analyzed by computer code to extract meaning, such as the identity of objects in the image and a deep understanding of the scene.

Data mining   This field, at the nexus of machine learning and statistics, finds techniques that allow very large amounts of data to be studied with the goal of extracting new discoveries. Astronomical sky surveys are the stereotypical example of big data that must be mined to extract discoveries regarding new asteroids or new planets from indirect data.

Eye tracking   A skill enabling a robot to visually examine the scene before it, identify the faces in the scene, mark the location of the eyes on each face, and then find the irises so that the gaze directions of the humans are known. Humans are particularly good at this even when we face other people at acute angles.

Hard AI   Also known as strong AI, this embodies the AI goal of going all the way toward human equivalence: matching natural intelligence along every possible axis so that artificial beings and natural humans are, at least from a cognitive point of view, indistinguishable.

Laser cutting   A rapid-prototyping technique in which flat material such as plastic or metal lays on a table and a high-power laser is able to rapidly cut a complex two-dimensional shape out of the raw material. Etching can also be performed by controlling the depth of the cut. Laser cutting affords very inexpensive and very fast creation of structural parts for robots.

LISP   A computer programming language that was popularized as a teaching tool in artificial intelligence classes across the country in the 1990s and as the main programming language for many early robots.

Manipulation   In robotics this denotes the challenge of interacting directly with objects in the physical world, often typically used to refer to the decision making and control needed to have robotic hands manipulate the common human world of cooking utensils and doorknobs, or the industrial world of assembly processes and rigs.

Micron   A thousandth of a millimeter. Typical red blood cells are 50 microns in diameter.

Robot   Never attempt to extract a definition for this word from a roboticist. Nearly all robotics researchers disagree completely regarding its meaning, and their definitions change rapidly with the onset of new innovations.

Singularity   This trend or event is hypothesized by futurists such as Raymond Kurzweil and Vernor Vinge as the oncoming rapid acceleration of AI capability to the point that it becomes self-reinforcing, with a runaway quality in which super-beings are rapidly created by a highly intelligent AI with or without the participation of humans (who may be plugged into the intelligence if we and AI systems hypothetically merge into a new race).

Snake robot   This refers to robotic systems that have very large numbers of motorized joints, enabling not only snakelike gaits such as the sidewinder, but often a number of gaits that natural snakes cannot exhibit, such as forming a vertical circle and moving as a vertically looping wheel.

Sonar   A very successful sensor on mobile robots of the 1970–1995 era in which a small, round transducer sends a sound wave out then measures the time it takes for an echo to return. Although we think of this as similar to echolocation in bats, their use of sonic wave patterns to fly, hunt prey and avoid one another is incomparably more complex than that of robot sonar.

Synapse   A junction between two nerve cells. Because each neuron in the brain may have up to 10,000 synaptic connections to other neurons, the number of synapses far exceeds the number of neurons in the brain: 100 trillion versus 100 billion respectively.

Telepresence   The ability to sense and act in a place removed from the actual location of one's body.

USAR   Urban search and rescue; an important subfield within robotics creating robots and robot interfaces that enable emergency responders to more effectively and less dangerously identify and rescue victims in unstructured disaster sites.

References

Apple Computer. 2011. "Siri. Your Wish Is Its Command." October 16. http://www.apple.com/iphone/features/siri.html (accessed May 9, 2012).

Au, S., M. Berniker, and H. Herr. 2008. Powered ankle-foot prosthesis to assist level-ground and stair-descent gaits. *Neural Networks (Special issue on Robotics and Neuroscience)* 21 (4) (May): 654–666.

Bartneck, C., M. Verbunt, O. Mubin, and A. Al-Mahmud. 2007. "To Kill a Mockingbird Robot." In *Proceedings of the 2nd ACM/IEEE International Conference on Human–Robot Interaction*. Washington, DC.

Bell, C., P. Shenoy, R. Chalodhorn, and R. Rao. 2008. Control of a humanoid robot by a noninvasive brain-computer interface in humans. *Journal of Neural Engineering* 5 (214): 214–220.

Bradski, G., and A. Kaehler. 2008. *Learning OpenCV: Computer Vision with the OpenCV Library*. Sebastopol, CA: O'Reilly Media.

Brown, Ben, Chris Bartley, Jennifer Cross, and Illah Nourbakhsh. 2012. "ChargeCar Community Conversions: Practical, Custom Electric Vehicles Now!" IEVC.

Brown, Ben, Garth Zeglin, and Illah Nourbakhsh. 2003. "Energy Storage Device Used in Locomotion Machine." US Patent 6,558,297.

Clark, H. H. 1996. *Using Language*. Cambridge: Cambridge University Press.

Clark, H. H., and S. E. Brennan. 1991. Grounding in Communication. In *Perspectives on Socially Shared Cognition*, ed. L. B. Resnick, R. M. Levine, and S. D. Teasley, 127–149. Washington, DC: American Psychological Association.

Corburn, Jason. 2005. *Street Science: Community Knowledge and Environmental Health Justice*. Cambridge, MA: MIT Press.

Deibert, R., J. Palfrey, R. Rohozinski, and J. Zittrain, eds. 2011. *Access Contested: Security, Identity, and Resistance in Asian Cyberspace*. Cambridge, MA: MIT Press.

DiSalvo, Carl, Marti Louw, Julina Coupland, and MaryAnn Steiner. 2009. Local Issues, Local Uses: Tools for Robotics and Sensing in Community Contexts. In *C & C '09: Proceedings of the ACM Conference on Creativity and Cognition*, 245–254. New York: ACM Press.

DiSalvo, C., I. Nourbakhsh, D. Holstius, A. Akin, and M. Louw. 2008. "The Neighborhood Networks Projects: A Case Study of Critical Engagement and Creative Expression through Participatory Design." In *Proceedings of the 2008 Participatory Design Conference*. Bloomington, IN.

Finn, P. 2011. "A Future for Drones: Automated Killing." *The Washington Post*. September 19.

Fong, T., I. Nourbakhsh, C. Kunz, L. Fluckiger, and J. Schreiner. 2005. "The Peer-to-Peer Human–Robot Interaction Project." In *Proceedings of AIAA Space*. Long Beach, CA.

Gieryn, Thomas. 1999. *Cultural Boundaries of Science*. Chicago: University of Chicago Press.

Gladwell, Malcolm. 2000. *The Tipping Point: How Little Things Can Make a Big Difference*. New York: Little Brown.

Hamner, E., T. Lauwers, D. Bernstein, I. Nourbakhsh, and C. DiSalvo. 2008. "Robot Diaries: Broadening Participation in the Computer Science Pipeline through Social Technical Exploration." In *Proceedings of the AAAI Spring Symposium on Using AI to Motivate Greater Participation in Computer Science*. Stanford, CA.

Holson, Laura. 2009. "Putting a Bolder Face on Google." *New York Times*. February 28.

Hooker, John. 2010. *Business Ethics as Rational Choice*. Upper Saddle River, NJ: Pearson Prentice-Hall.

Jackson, J. 2007. Microsoft robotics studio: A technical introduction. *IEEE Robotics & Automation Magazine* 14 (4): 82–87.

Kahn, P., N. Freier, T. Kanda, H. Ishiguro, J. Ruckert, R. Severson, and S. Kane. 2008. "Design Patterns for Sociality in HumanRobot Interaction." In *Proceedings of Human–Robot Interaction*. New York: ACM Press.

Kazerooni, Homayoon. 2012. Berkeley Robotics & Human Engineering Laboratory. http://bleex.me.berkeley.edu/ (accessed May 9, 2012).

Kelly, Kevin. 2010. *What Technology Wants*. New York: Viking Press.

Kurzweil, Ray. 2006. *The Singularity Is Near: When Humans Transcend Biology*. New York: Penguin Group.

Lerner, Steve. 2010. *Sacrifice Zones*. Cambridge, MA: MIT Press.

Leveson, N. G., and C. S. Turner. 1993. An investigation of the Therac 25 accidents. *Computer* 26 (7): 18–41.

Lewis, M., S. Carpin, and S. Balakirsky. 2009. "Virtual Robots RoboCup-pRescue Competition: Contributions to Infrastructure and Science." In *Proceedings of IJCAI Workshop on Competitions in Artificial Intelligence and Robotics*.

Lewis, M., and K. Sycara. 2011. "Network-Centric Control for Multirobot Teams in Urban Search and Rescue." In *Proceedings of the 44th Hawaiian International Conference on Systems Sciences*.

Linder, T., V. Tretyakov, S. Blumenthal, P. Molitor, D. Holz, R. Murphy, S. Tadokoro, and H. Surmann. 2010. "Rescue Robots at the Collapse of the Municipal Archive of Cologne City: A Field Report." In *IEEE International Workshop on Safety Security and Rescue Robotics (SSRR)*.

McGeer, T. 1990. Passive dynamic walking. *International Journal of Robotics Research* 9 (2) (April): 62–82.

Mellinger, Daniel, Nathan Michael, and Vijay Kumar. 2010. "Trajectory Generation and Control for Precise Aggressive Maneuvers with Quadrotors." In *Proceedings of the International Symposium on Experimental Robotics*. New Delhi, India.

Moravec, Hans. 1990. *Mind Children: The Future of Robot and Human Intelligence*. Cambridge, MA: Harvard University Press.

Müller, Jörg, Juliane Exeler, Markus Buzeck, and Antonio Krüger. 2009. ReflectiveSigns: Digital signs that adapt to audience attention. *Pervasive Computing: Lecture Notes in Computer Science* 5538: 17–24.

Murphy, M., S. Kim, and M. Sitti. 2009. Enhanced adhesion by gecko inspired hierarchical adhesives. *ACS Applied Materials and Interfaces* 1 (4): 849–855.

Nilsson, N. 1984. *Shakey the Robot*. SRI International Technical Note No. 323.

Nourbakhsh, I., C. Acedo, R. Sargent, C. Strebel, L. Tomokiyo, and C. Belalcazar. 2010. "GigaPan Conversations: Diversity and Inclusion in the Community." In *Proceedings of the International Scientific Conference on Technology for Development*, 53–60. Lausanne, Switzerland: United Nations.

Nourbakhsh, Illah, David Andre, Carlo Tomasi, and Michael Genesereth. 1997. Mobile robot obstacle avoidance via depth from focus. *Robotics and Autonomous Systems* 22:151–158.

Nourbakhsh, Illah, Judith Bobenage, Sebastien Grange, Ron Lutz, Roland Meyer, and Alvaro Soto. 1999. An affective mobile educator with a full-time job. *Artificial Intelligence* 114 (1–2): 95–124.

Nourbakhsh, I., E. Hamner, D. Bernstein, K. Crowley, E. Ayoob, M. Lotter, S. Shelly, et al. 2006. The personal exploration rover: Educational assessment of a robotic exhibit for informal learning venues. *International Journal of Engineering Education, Special Issue on Trends in Robotics Education* 22 (4): 777–791.

Nourbakhsh, Illah, Rob Powers, and Stan Birchfield. 1995. Dervish: An office-navigating robot. *AI Magazine* 16 (2): 53–60.

Omer, A. M. M., R. Ghorbani, Hun-ok Lim, and A. Takanishi. 2009. "Semi-Passive Dynamic Walking for Biped Walking Robot Using Controllable Joint Stiffness Based on Dynamic Simulation." In *Advanced Intelligent Mechatronics*, 2009. Singapore.

Perez, Sarah. 2011. "Euclid Elements Emerges from Stealth, Debuts 'Google Analytics for the Real World.'" *techcrunch.com*. November 3. http://techcrunch.com/2011/11/03/euclid-elements-emerges-from-stealth-debuts-google-analytics-for-the-real-world/ (accessed May 9, 2012).

Power, D. J., ed. 2002. Beer and Diapers. *DSS News* 3 (23) (November 10). http://www.dssresources.com/newsletters/66.php (accessed May 9, 2012).

Quigley, M., B. Gerkey, K. Conley, J. Faust, T. Foote, J. Leibs, E. Berger, R. Wheeler, and A. Ng. 2009. "ROS: An Open-Source Robot Operating System." In *Proceedings of IEEE ICRA 2009*. Kobe, Japan.

Rowe, Anthony, Charles Rosenberg, and Illah Nourbakhsh. 2002. *A Low Cost Embedded Color Vision System*. Lausanne, Switzerland: IROS.

Shropshire, Corilyn. 2006. "Fast-Food Assistant 'Hyperactive Bob' Example of Robots' Growing Role." *Pittsburgh Post-Gazette*. June 16.

Singer, P. W. 2009. *Wired for War: The Robotics Revolution and Conflict in the 21st Century*. New York: Penguin Books.

Srinivasa, S., D. Ferguson, C. Helfrich, D. Bernson, A. Coilet, R. Diankov, G. Gallagher, G. Hollinger, J. Kuffner, and M. Vande Weghe. 2010. Herb: A home exploring robotic butler. *Autonomous Robots Journal* 28:5–20.

State of Nevada. 2011. Assembly Bill No. 511. Section 8. Committee on Transportation.

Steinfeld, A., T. Fong, D. Kaber, M. Lewis, J. Scholtz, A. Schultz, and M. Goodrich. 2006. "Common Metrics for Human–Robot Interaction." In *Proceedings of Human–Robot Interaction*. Salt Lake City, UT.

A Swarm of Nano Quadrotors. 2012. http://www.youtube.com/watch ?v=YQIMGV5vtd4 (accessed January 31, 2012).

Turkle, Sherry. 2011. *Alone Together: Why We Expect More from Technology and Less from Each Other*. New York: Basic Books.

Walker, Matt. 2009. "Ant Mega-Colony Takes over World." *BBC Earth News*. July 1.

Wilber, B. M. 1972. "A Shakey Primer." Technical Report. Stanford Research Institute, Menlo Park, CA. November.

Index